# Sexual and Asexual Reproduction in Forest Plants

# Sexual and Asexual Reproduction in Forest Plants

Guest Editor

**Ling Yang**

MDPI

Basel • Beijing • Wuhan • Barcelona • Belgrade • Novi Sad • Cluj • Manchester

*Guest Editor*
Ling Yang
College of Forestry
Beijing Forestry University
Beijing
China

*Editorial Office*
MDPI AG
Grosspeteranlage 5
4052 Basel, Switzerland

This is a reprint of the Special Issue, published open access by the journal *Plants* (ISSN 2223-7747), freely accessible at: https://www.mdpi.com/journal/plants/special_issues/9PT90IDF2S.

For citation purposes, cite each article independently as indicated on the article page online and as indicated below:

| Lastname, A.A.; Lastname, B.B. Article Title. *Journal Name* **Year**, *Volume Number*, Page Range. |
|---|

**ISBN 978-3-7258-7544-3 (Hbk)**
**ISBN 978-3-7258-7545-0 (PDF)**
**https://doi.org/10.3390/books978-3-7258-7545-0**

# Contents

# About the Editor

**Ling Yang**

Ling Yang is a Professor and Doctoral Supervisor at the College of Forestry, Beijing Forestry University. She earned her PhD in Forest Bioengineering from Northeast Forestry University, completed her postdoctoral research in Biology there, and served as a visiting scholar at Lomonosov Moscow State University and the Russian Academy of Sciences' Institute of Plant Physiology.

Her research focuses on the field of sexual and asexual reproduction in forest plants, specifically encompassing core areas such as somatic embryogenesis, stem cell culture, cryopreservation, stress resistance physiology, nitric oxide signaling, seed dormancy, and multi-varietal forestry. She has successfully established large-scale cell engineering propagation systems for over ten economic and timber tree species (e.g., Pinus koraiensis, Fraxinus mandshurica, Sorbus pohuashanensis) and developed tree breeding technologies and plant stem cell cryopreservation banks through international cooperation. As a leading researcher, she has presided over 30 projects, including National Key R&D Program tasks, National Natural Science Foundation projects, and international cooperation initiatives, secured 17 authorized patents, and formulated 2 local standards. She has published more than 150 academic papers (nearly 60 indexed in SCI) and authored 8 monographs and 5 textbooks.

In terms of teaching and talent cultivation, she has supervised over 60 master's and doctoral students (including international joint-training candidates) and has received 9 provincial and municipal awards. Her research integrates basic theory with practical applications, making significant contributions to the understanding of forest plant reproductive mechanisms, germplasm conservation, and global sustainable forestry.

# Preface

It is our pleasure to present this Reprint, a focused collection on sexual and asexual reproduction in forest plants—core to biodiversity conservation, ecological resilience, and sustainable forestry. As Guest Editors, we have curated it to address urgent forest challenges amid climate change and resource demands, bridging fundamental insights with practical applications.

Covering woody oilseed trees, ornamental shrubs, clonal dioecious plants, and more, the Reprint synthesizes reproductive mechanisms (physiological, genetic, molecular) and innovations like in vitro regeneration, cryopreservation, and pathogen control. It serves researchers, forest managers, breeders, and policymakers, offering evidence-based strategies for restoration, propagation, and genetic diversity protection.

With appreciation of the scientific rigor of the authors, *Plants* editorial team, and reviewers, we hope this Reprint inspires further research and collaboration to sustain forests through an enhanced understanding of plant reproductive biology.

**Ling Yang**
*Guest Editor*

*plants*

MDPI

*Article*

# Colchicine-Induced Tetraploidy in Protocorms of *Aerides rosea* Lodd. ex Lindl. and Paxton. and Its Identification

**Li Wang [1,2,3,†], Pengrui Zheng [1,2,†], Hong Ge [1,2], Xin Zhao [1,2], Yaping Kou [1,2], Shuhua Yang [1,2], Xiaonan Yu [3,*] and Ruidong Jia [1,2,*]**

1 State Key Laboratory of Vegetable Biobreeding, Institute of Vegetables and Flowers, Chinese Academy of Agricultural Sciences, Beijing 100081, China; lilili0126@163.com (L.W.); zhengpengrui326@163.com (P.Z.); gehong@caas.cn (H.G.); zhaoxin01@caas.cn (X.Z.); kouyaping@caas.cn (Y.K.); yangshuhua@caas.cn (S.Y.)

2 Key Laboratory of Biology and Genetic Improvement of Flower Crops (North China), Ministry of Agriculture and Rural Affairs, Beijing 100081, China

3 School of Landscape Architecture, Beijing Forestry University, Beijing 100083, China

* Correspondence: yuxiaonan626@126.com (X.Y.); jiaruidong@caas.cn (R.J.); Tel.: +86-010-62338304 (X.Y.); +86-10-8210-9542 (R.J.)

† These authors contributed equally to this work.

**Abstract:** *Aerides rosea* (Orchidaceae) boasts high ornamental value due to its pleasant aroma, foxtail spike, and elegant floral morphology. Inducing *A. rosea* to become tetraploid enhances horticultural traits and facilitates fertile intergeneric hybrids through crosses with other market-available tetraploid species. The experimental design involved the application of colchicine at varying concentrations—0.05%, 0.1%, and 0.2%—to a solid medium. Exposure durations were 5, 10, and 15 days, with treatments conducted under sterile conditions on 6-week-old protocorms post-germination. Results indicated that the protocorms were sensitive to colchicine concentrations exceeding 0.05%, with high concentrations leading to a mortality rate exceeding 50%. Flow cytometry (FCM) with 4′,6-diamidino-2-phenylindole (DAPI) staining confirmed a doubling of chromosome numbers in tetraploid plants ($2n = 4x = 76$) compared to diploid controls ($2n = 2x = 38$). Induction efficiency was significantly influenced by colchicine concentration and treatment duration. A 10-day treatment with 0.2% colchicine yielded a 70.00% tetraploid induction rate; however, considering protocorm survival, a 5-day treatment with 0.05% colchicine was preferable, achieving a 63.55% survival rate and a 56.67% tetraploid induction rate. Tetraploid plants exhibited distinct morphological traits, such as a more compact growth habit, thicker leaves, and increased stem and root thickness. Leaf morphology changes included larger stomata with reduced density, denser spongy mesophyll, and more pronounced venation. Tetraploids also demonstrated a 1.94-fold increase in genome size compared to diploids. The tetraploid genotypes developed in this study hold significant potential for future *Aerides* breeding programs.

**Keywords:** *Aerides*; chromosome doubling; orchid breeding; polyploid

## 1. Introduction

Orchidaceae, with approximately 30,000 species, is one of the largest families of flowering plants, exhibiting tremendous floral diversity [1]. Orchids are closely associated with forests, as many species primarily grow within forest ecosystems, where they play essential roles [2]. For hundreds of years, orchids have been prized by collectors and adventurers. Some of them, such as *Phalaenopsis* and *Cattleya*, have been cultivated as major ornamental crops. However, many species with significant horticultural value remain underutilized. *Aerides rosea*, among the vast array of orchid species, stands out as a particularly noteworthy example of the family's diversity and ecological significance. This species has not been fully exploited for its horticultural potential. This rare and endangered [3] epiphytic orchid (living on trunks or branches of the host trees without absorbing nutrients and water from the hosts) is primarily distributed in the forests of southwestern China, Bhutan, northeastern India, Myanmar, Laos, and Vietnam [4]. Characterized by monopodial branching and succulent leaves, *A. rosea* produces long inflorescences densely packed with numerous flowers. The flowers are pink, pleasantly fragrant, and uniquely shaped, making this species highly valuable as an ornamental plant for both potted and cut flower markets [5]. As a member of the subtribe Aeridinae, *A. rosea* can hybridize with species from related genera within the subtribe. Its fragrant flowers make it a valuable resource for the production of many scented artificial hybrids and cultivars [6]. Within the subtribe Aeridinae, *Phalaenopsis* and *Vanda* are among the most popular and sought-after orchids in the global market [7]. According to the RHS database [8], there are 469 hybrid offspring involving *Aerides*, with 286 intergeneric hybrids with *Vanda* and 16 with *Phalaenopsis*. This indicates the extensive use of *Aerides* in intergeneric hybridization.

However, achieving successful hybridization can be challenging. Interspecific hybrids often suffer from reduced fertility due to problems with chromosome pairing and segregation during meiosis [9]. Polyploid breeding can address this issue of hybrid infertility by restoring fertility. Additionally, it is crucial to overcome barriers to hybridization. In some cases, obtaining desirable crosses is challenging due to differences in ploidy levels between potential parents. By adjusting the ploidy levels to be compatible before hybridization, these interploidy barriers can be overcome, allowing successful hybridization. [10]. Polyploids are defined as individuals with three or more sets of chromosomes in their somatic cells [11]. They are valuable to breeders due to their potential traits, including increased leaf thickness and length, larger and more textured flowers, prolonged blooming periods, thicker stems, greater biomass, and improved stress resistance compared to their diploid counterparts [12,13]. Polyploidy has been an important factor in the development of improved commercial varieties and hybrids of orchids [14]. However, there are currently no reports of tetraploid germplasm of *Aerides* being used for breeding.

Polyploid induction is an important method for genetic improvement and hybridization in orchids [15]. The primary mechanisms of natural polyploidy in orchids include endopolyploidy and hybridization [16]. Artificial induction is mainly achieved by applying antimitotic agents to tissues, organs, or whole plants [17]. Studies have shown that the efficiency of artificial polyploid induction in Orchidaceae depends on concentration, exposure time of the antimitotic agent, the type and age of explants, and the in vitro induction protocol [18]. The most widely used antimitotic agent for polyploid induction in plants is colchicine [19,20]. Significant progress has been made in inducing polyploid orchids in vitro using colchicine. This progress includes the successful somatic chromosome doubling in species such as *Phalaenopsis* [21], *Vanda* [22], *Cattleya* [23], *Cymbidium* [24], and *Dendrobium* [25], achieved through the application of colchicine in both solid and liquid media at concentrations ranging from 0.005% to 0.5%. It inhibits the formation of spindle fiber and effectively arrests mitosis at the anaphase stage. At this point, the chromosomes are doubled after cells pass another cycle. Orchid polyploidy can be induced either by culturing explants on a medium with colchicine at a precise concentration [21] or by submerging them in a colchicine solution [26]. Nonetheless, there are no reports on the polyploidization of *A. rosea*.

In addition to an effective system for polyploidy induction, confirmation methods are crucial for obtaining polyploids. Flow cytometry (FCM) is a rapid and relatively accurate method for the large-scale identification of artificially induced polyploids [27]. Chromosome counting in mitotic cells of root tips is an accurate procedure to determine ploidy, but it is time consuming and requires significant expertise. Therefore, attempts have been made to find indirect methods for ploidy determination. The area of stomata has been used to differentiate diploid and tetraploid regenerants of orchids [15]. The number of chloroplasts per guard cell pair [28] and differences in stomata density [29] have also been used as criteria in the distinction between diploid and tetraploid plants.

Inducing tetraploidy in *A. rosea* not only improves its horticultural traits but also enables crosses with other market-available tetraploid species, resulting in fertile intergeneric hybrids. In this study, we achieved tetraploids in *A. rosea* through in vitro colchicine treatment of protocorms. The resultant polyploid plants were identified based on morphological characteristics and confirmed via FCM and chromosome counts in root tip mitotic cells.

## 2. Materials and Methods

### 2.1. Plant Material and In Vitro Protocorm Establishment

*A. rosea* plants (2n = 2x = 38) were cultivated in the greenhouse of the Institute of Vegetables and Flowers, Chinese Academy of Agricultural Sciences, located in Beijing, China. The plants were hand pollinated at full bloom in early June 2021, and capsules were harvested before dehiscence in late September. Detailed phenotypic characteristics are shown in Figure 1.

**Figure 1.** Morphology of *A. rosea* flowers, pods, and protocorms. (**A**) Flowers; (**B**) fruit pods; and (**C**) 6-week-old protocorms. Bars = 1 cm.

To produce in vitro protocorms, the capsules were disinfected in 75% ethanol for 30 s, washed once with sterile water, then were surface disinfected in a 10% (*v*/*v*) NaClO solution with an active chlorine content of 6–14% for 15 min. After rinsing three times with sterile distilled water, the capsules were cut with a sterile blade. Using sterilized tweezers, the capsule was held at one end above the mouth of the culture bottle, and carefully, the seeds were scraped out from the capsule into the culture bottle using the tip of the blade. The seeds were then spread onto the surface of the 1/2 Murashige and Skoog (MS) culture medium [30] with 5 $g \cdot L^{-1}$ carrageenan [31], 20 $g \cdot L^{-1}$ sucrose, and 100 $mL \cdot L^{-1}$ coconut water [32] at pH 5.8. The cultures were incubated at 25 °C under a 16/8 h light/dark photoperiod at a light intensity of 40 $\mu mol/m^2/s$ provided by white florescent tubes. Protocorms developed 6 weeks after sowing. Subsequently, protocorms with shoot apical meristem were used as explants for polyploid induction (Figure 1C).

### 2.2. Polyploidy Induction

Six-week-old protocorms of *A. rosea* were used for colchicine treatment. The protocorms used in the study were approximately 1–2 mm in diameter. They were cultured in the 1/2 MS culture medium with 1.0 $mg \cdot L^{-1}$ 6-benzylaminopurine, 5 $g \cdot L^{-1}$ carrageenan, 20 $g \cdot L^{-1}$ sucrose, 1.5 $g \cdot L^{-1}$ activated carbon, and 100 $mL \cdot L^{-1}$ coconut water, supplemented with colchicine at concentrations of 0, 0.05%, 0.1%, and 0.2% for 5, 10, and 15 days

(Table 1). The colchicine solutions were sterilized by passing them through a 0.45 μm Millipore filter [21]. The experiment followed a completely randomized design with a factorial arrangement. Fifteen protocorms were used per replication, and three replicates were established for each colchicine treatment. After exposure to colchicine, the protocorms were rinsed three times using sterile distilled water and transferred onto fresh antimitotic-free seedling formation medium (growth conditions as mentioned above). The surviving protocorms were subcultured in medium and developed into plantlets within two months. Survival rates of protocorms were recorded at 6 weeks after colchicine treatment. The surviving protocorms were then transferred to a similar medium for further ploidy level analysis.

**Table 1.** Effects of colchicine on the survival rate and the mutation rate of *A. rosea* protocorms.

| Antimitotic Agent | Treatment | Exposure Time (d) | Concentration % | Survival Rate % | Rate of Mutation % |
|---|---|---|---|---|---|
| Control | CK | 0 | 0 | 98.56 ± 3.02 a | 0 |
| Colchicine | 1 | 5 | 0.05 | 63.55 ± 6.08 b | 56.67 ± 2.89 c |
| | 2 | 10 | 0.05 | 47.61 ± 2.09 c | 41.67 ± 2.89 e |
| | 3 | 15 | 0.05 | 16.93 ± 1.7 f | 41.67 ± 5.77 e |
| | 4 | 5 | 0.1 | 46.09 ± 0.61 c | 43.33 ± 2.89 e |
| | 5 | 10 | 0.1 | 33.06 ± 1.69 de | 46.67 ± 2.89 de |
| | 6 | 15 | 0.1 | 10.24 ± 1.04 g | 51.67 ± 2.89 cd |
| | 7 | 5 | 0.2 | 34.92 ± 1.37 d | 63.33 ± 2.89 b |
| | 8 | 10 | 0.2 | 29.74 ± 1.37 c | 70.00 ± 5.00 a |
| | 9 | 15 | 0.2 | 6.69 ± 0.45 g | 51.67 ± 2.89 cd |

Note: Number of treated protocorms = 45. Data are the means ± SE of three replicates. Different lowercase letters in the same column indicate significant differences ($p < 0.05$).

### 2.3. Flow Cytometry Polyploidy Analysis

Fresh leaves of 7-month-old seedlings were harvested after colchicine treatments for ploidy level identification using a CyFlow ploidy analyzer (Sysmex-Partec GmbH, Norderstedt, Germany). Sample preparation was performed according to the CyStain UV Precise P kit's (Sysmex-Partec GmbH, Norderstedt, Germany) manual with minor modifications as follows: Samples were chopped in 400 μL nuclear extraction buffer for 30 s using a razor blade. The nuclei were mixed with 1600 μL of 4′,6-diamidino-2-phenylindole (DAPI) staining buffer and filtered through a 30 μm Cell-Trics filter (Sysmex-Partec GmbH, Norderstedt, Germany) into a sample tube. For each experimental setup, samples of young leaves from the in vitro seedlings of untreated (control) diploid *A. rosea* were selected as the calibration standard to adjust the gain of the flow cytometer such that the first G0/G1 peaks were set at channel 100 on a 1000 channel scale. Histograms were generated after analyzing ≥5000 nuclei using the De Novo FCS Express 7 software (De Novo Software, Los Angeles, CA, USA). The number of polyploid plants was recorded based on FCM analysis.

### 2.4. Morphological Analysis

Fifteen diploid and fifteen tetraploid plantlets were cultured in a plant tissue culture laboratory for 24 months. Morphological characteristics, including plant height, stem diameter, leaf count, maximum leaf length and width, leaf thickness, root count, and maximum root diameter, were analyzed. Leaf and root counts were directly recorded, while other parameters were measured with a digital vernier caliper.

### 2.5. Stomatal Observation and Leaf Anatomy

Leaves from diploid and colchicine-treated tetraploid plants were selected. Three leaves from each plant were cut from the middle regions of the third fully expanded leaves from the top of the stem. To observe stomata characteristics, upper and lower epidermal peels were taken from the leaves using forceps. Each peel was mounted on a glass slide with a drop of water and then covered with a cover slip [33]. The slides were observed under a microscope (OLYMPUS BX53, Olympus Corporation, Tokyo, Japan). The stomatal length and width of 10 randomly selected stomata from each leaf were measured. Then, stomatal density was recorded in two randomly selected microscopic fields of view on the adaxial and abaxial sides of each leaf.

Diploid and tetraploid leaf samples (5 mm × 5 mm) were cut from the middle regions of the third fully expanded leaves from the top of the stem and fixed in FAA (5% formaldehyde: 5% acetic acid: 90% ethanol, *v*/*v*) at room temperature overnight. Dehydration was performed in a series of ethanol solutions (50%, 70%, 85%, 95%, 100%, 100% *v*/*v*) for 15 min each. The samples were then processed through a 1:1 xylene: ethanol (*v*/*v*) mixture, followed by two rounds of xylene clearing for 15 min each. Gradual infiltration with paraffin wax at 62 °C was performed, involving a 1:1 xylene: paraffin (*v*/*v*) mixture and three rounds of paraffin immersion for 4 h each. The samples were cut into 10 μm transverse sections using a semi-motorized rotary microtome (Leica RM 2255, Leica Biosystems, Nussloch, Germany) and stained with toluidine blue. Sections were observed under an optical microscope (OLYMPUS BX53, Olympus Corporation, Tokyo, Japan). Four samples were analyzed for each ploidy level. For each sample, at least three random cross sections were chosen to measure leaf, mesophyll, vein, and upper and lower epidermal thicknesses.

### *2.6. Chromosome Counting*

Root tips approximately 0.5 cm in length were collected and pretreated with a 0.1% colchicine solution for 6 h at 20 °C in the dark to induce chromosome contraction. Following this, the root tips were fixed overnight in Carnoy's solution (ethanol:glacial acetic acid = 3:1). The fixed samples were then washed twice with distilled water and dehydrated in a series of ethanol solutions of varying concentrations, ultimately being preserved in 75% ethanol at 4C. Cell wall separation was facilitated by immersing the root tips in 0.075 mol/L KCl for 30–60 min at 37 °C. Subsequently, the root tips were treated with 1 mol/L HCl and heated in a water bath at 60 °C for 10 min. The samples were then squashed in a modified phenol fuchsin solution and observed under a microscope (OLYMPUS BX53, Olympus Corporation, Tokyo, Japan) to determine the chromosome number.

### *2.7. Genome Size Measurement*

The nuclear DNA content of three plantlets from each solid diploid and tetraploid were estimated using a CyFlow ploidy analyzer with a 532 nm laser light source. Due to the unknown genome size of diploid *A. rosea* and the extensive endopolyploidy in orchids, making them potentially inaccurate as reference standards [34], *Zea mays* L. 'B73' [35] was chosen as the internal reference standard. Fresh leaves from the samples and the internal reference standard of *Z mays* 'B73' were simultaneously chopped with a sharp razor blade according to the manual for the Partec Cystain Propidium Iodide (PI) Absolute P kit (Sysmex-Partec GmbH, Norderstedt, Germany). Histograms were generated after analyzing ≥2000 nuclei [36], and representative flow histograms were analyzed using the De Novo FCS Express 7 software.

Values for the nuclear DNA content of the test sample were estimated using the following formula [37]:

$$\text{Sample 2C DNA content} = [(\text{sample G1 peak mean})/(\text{standard G1 peak mean})] \times \text{standard 2C DNA content (pg DNA)}$$

$$1 \text{ pg DNA} = 0.978 \times 109 \text{ bp}$$

### *2.8. Statistical Analysis*

Plant height was measured as the vertical distance from the base of the plant to the highest point of the plant; shoot diameter was measured across the root to the nearest millimeter scale; root length was measured from the leaf base to the tip of the roots to the nearest millimeter scale; root diameter was measured across the root to the nearest millimeter scale; leaf thickness was measured as the distance between the two surfaces at the mid-point of the leaf; leaf length was measured from the leaf base to the tip to the nearest millimeter scale; leaf width was measured across the leaf to the nearest millimeter scale; the total number of roots of each plant was counted and recorded; and the total number of leaves of each plant was counted and recorded. The data analysis was conducted using Excel 2016 for basic analysis. One-way ANOVA and mean comparisons among treatments using the LSD method (at alpha = 0.05) were performed using SPSS 22. The experiment was biologically replicated three times. The values in the table represent the range of data changes (±) with respect to the standard deviation.

## 3. Results

### 3.1. The Impact of Colchicine Treatment on Survival and Induction Rates

The different concentrations of colchicine (0.05%, 0.1%, and 0.2%) and varying exposure times (5 days, 10 days, and 15 days) significantly affect the survival and mutation rates of protocorms. Data indicate that when colchicine concentration exceeds 0.05%, protocorm survival rate drops significantly. For instance, at 0.05% concentration, the survival rate is 63.55% after 5 days of treatment, but it decreases to 16.93% after 15 days. Similarly, at 0.2% concentration, the survival rate is 34.92% after 5 days and drops to 6.69% after 15 days. These results show that higher concentrations of colchicine have strong toxicity to protocorms, leading to high mortality rates. Additionally, high concentrations (0.2%) of colchicine result in mortality rates exceeding 50% across all exposure times, with a mortality rate of 93.31% after 15 days of treatment (Table 1).

The interaction between colchicine concentration and exposure duration was crucial for the induction rate, highlighting the importance of optimizing both factors for in vitro tetraploidy induction in *A. rosea*. The highest survival rate, 70.00%, occurred in the 0.2% treatment for 10 days, although the actual survival rate was 29.74%. Compared to Treatment 8, which had the highest induction rate, Treatment 7 used the same concentration but had a shorter 5-day treatment duration. This led to a 6.33% decrease in tetraploid induction rate and a 5.18% drop in protocorm survival rate. Treatment 1, with a 0.05% concentration for 5 days, experienced a 13.33% reduction in induction rate but a 33.81% increase in protocorm survival rate.

Tetraploid plants were identified using FCM on 121 surviving regenerated plants (Table 1). Diploid controls showed an initial G1 peak at channel 100 (Figure 2A), whereas a peak at channel 200 indicated tetraploidy (Figure 2B). All control plants were confirmed as diploid. Histograms in Figure 2 revealed three distinct peaks for both diploid and tetraploid samples, suggesting endopolyploidy in in vitro-regenerated cells. Chromosome counts confirmed the FCM results: diploids had 2n = 2x = 38 (Figure 2C), and tetraploids had 2n = 4x = 76 (Figure 2D).

### 3.2. Morphological Comparison

Twenty-four months after colchicine treatment, tetraploid plants (Figure 3B–I) exhibited significant morphological differences compared to diploid plants (Figure 3A). These differences included shorter stature, more compact growth, thicker stems and roots, rougher leaf texture, and thicker leaves. Under the same in vitro conditions, significant morphological differences were observed between diploid and tetraploid *A. rosea* plants (Table 2).

The average leaf width was 6.21 mm for diploids and 8.34 mm for tetraploids. Leaf thickness increased from 0.97 mm in diploids to 1.42 mm in tetraploids. Tetraploid plants had larger root diameters, measuring at 4.30 mm compared to 2.29 mm in diploids. Stem diameters also increased, from 2.01 mm in diploids to 3.22 mm in tetraploids. However, tetraploid plants were shorter and had shorter leaves than diploids. There was no significant difference in the number of leaves and roots between the two ploidy levels.

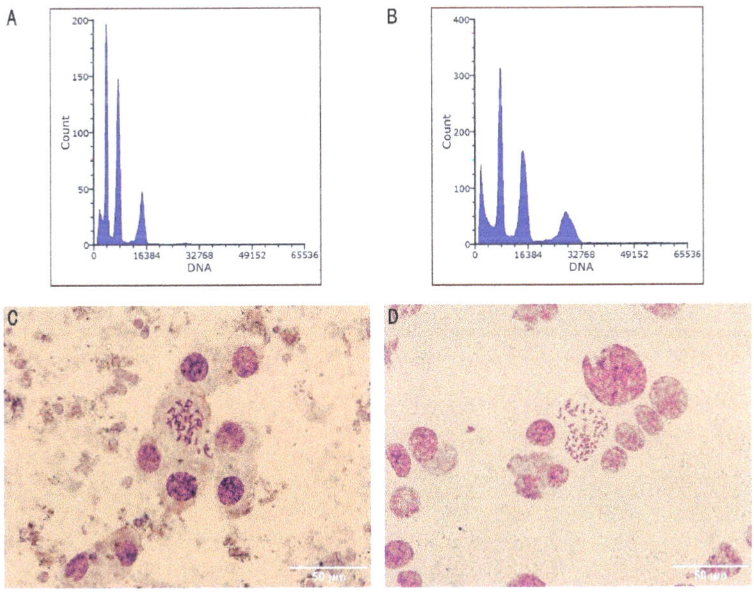


**Figure 2.** Ploidy detection of diploid and tetraploid plantlets via FCM and chromosomal counting. FCM histograms of DAPI-stained nuclei from in vitro leaf tissue for diploid (**A**) and tetraploid plantlets (**B**); chromosome of root tip cells from diploid 2n = 2x = 38 (**C**); and tetraploid 2n = 4x = 76 (**D**). Bars = 50 μm.

**Figure 3.** Morphologies of one-year-old diploid and tetraploid plantlets after colchicine treatment. (**A**) Diploid control plantlet; (**B**) 0.05% colchicine; (**C**) 0.05% colchicine; (**D**) 0.05% colchicine; (**E**) 0.05% colchicine; (**F**) 0.05% colchicine; (**G**) 0.1% colchicine; (**H**) 0.1% colchicine; and (**I**) 0.2% colchicine. Bars = 1 cm.

**Table 2.** Comparison of morphological indexes of diploids and tetraploids induced by colchicine.

| Phenotypic Characteristics | Diploid | Tetraploid |
|---|---|---|
| Plant height (mm) | 49.77 ± 1.85 a | 29.75 ± 1.04 b |
| Stem diameter (mm) | 2.01 ± 0.03 b | 3.22 ± 0.02 a |
| Number of leaves per plant | 5.25 ± 0.5 a | 4.25 ± 0.96 a |
| Leaf length (mm) | 21.03 ± 1.04 a | 11.13 ± 0.34 b |
| Leaf width (mm) | 6.21 ± 0.03 b | 8.34 ± 0.02 a |
| Aspect ratio of leaves | 3.77 ± 0.24 a | 1.70 ± 0.07 b |
| Leaf thickness (mm) | 0.97 ± 0.14 b | 1.42 ± 0.13 a |
| Root length (mm) | 43.21 ± 4.89 b | 26.82 ± 2.29 a |
| Root diameter (mm) | 2.29 ± 0.18 b | 4.30 ± 0.36 a |
| Number of roots per plant | 5.25 ± 0.50 a | 4.25 ± 0.96 a |

Note: Data are the means ± SE of three replicates. Different lowercase letters in the same column indicate significant differences ($p < 0.05$).

### *3.3. Stomatal and Mesophyll Cell Differences*

Differences in stomata between tetraploid and diploid plants provide a straightforward method for identifying polyploids. The stomatal density in tetraploids (Figure 4B) is significantly lower than in diploids (Figure 4A), with diploids at 31.09 μm and tetraploids at 8.06 μm. Tetraploid stomata (Figure 4D) are larger, with lengths and widths 39.51% and 52.86% greater than those of diploid stomata (Figure 4C) (Table 3). Both diploid (Figure 4E) and tetraploid (Figure 4F) plants have polygonal epidermal cells.

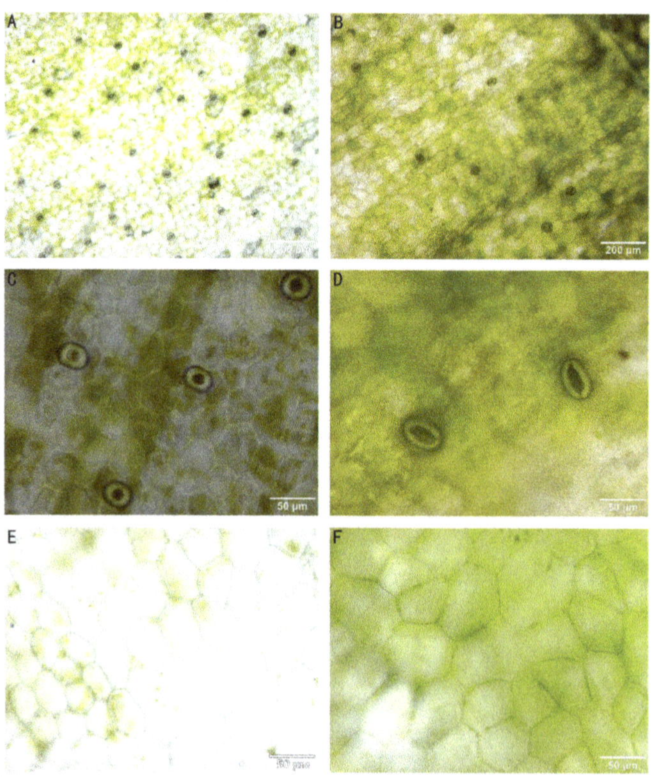


**Figure 4.** Leaf epidermal characteristics of diploids (**left**) and tetraploids (**right**). Abaxial epidermal stomatal density (**A**,**B**); stomatal size (**C**,**D**); and adaxial epidermal cell densities (**E**,**F**).

**Table 3.** Leaf anatomical structure of diploid and tetraploid plantlets.

| Stomatal Properties | Diploid | Tetraploid |
|---|---|---|
| Stomatal width (μm) | 32.90 ± 1.83 b | 45.90 ± 4.52 a |
| Stomatal height (μm) | 30.02 ± 0.84 b | 45.89 ± 5.86 a |
| Stomatal aspect ratio | 1.10 ± 0.06 a | 1.02 ± 0.18 a |
| Stomatal density (mm$^2$) | 31.09 ± 1.35 a | 8.06 ± 1.12 b |
| Leaf vein diameter (μm) | 62.43 ± 7.80 b | 91.47 ± 12.91 a |
| Upper epidermal thickness (μm) | 43.29 ± 3.43 b | 54.81 ± 3.56 a |
| Lower epidermal thickness (μm) | 30.44 ± 0.90 b | 47.76 ± 3.58 a |
| Mesophyll thickness (mm) | 0.83 ± 0.01 b | 1.04 ± 0.05 a |

Note: Data are the means ± SE of three replicates. Different lowercase letters in the same column indicate significant differences ($p < 0.05$).

Longitudinal sections of the leaves show that both diploid (Figure 5A) and tetraploid (Figure 5B) orchid leaves have uniform mesophyll cells without distinct palisade or spongy tissues. The mesophyll cells are tightly packed and vary in shape from polygonal to elliptical. Stomata are distributed on the lower epidermis of the leaves. Table 3 shows that the thickness of epidermal cells, veins, and mesophyll in tetraploid leaves is significantly greater than that in diploid leaves.

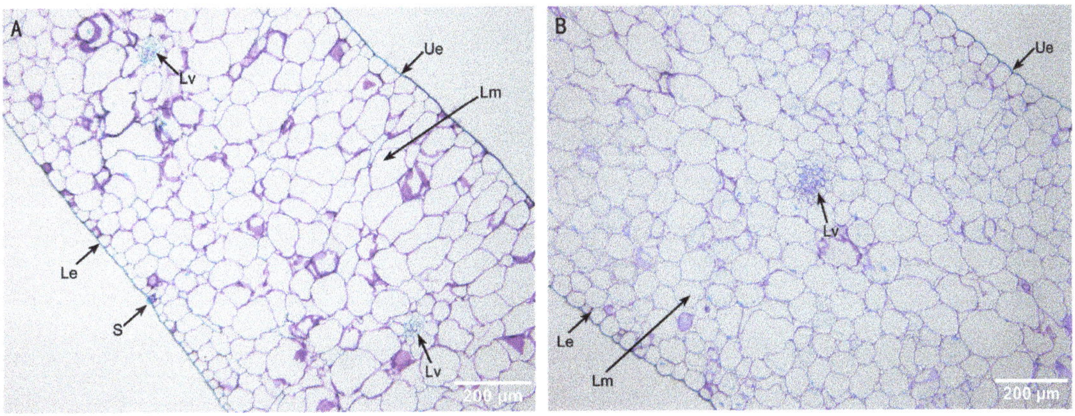


**Figure 5.** Leaf cross-section structure of diploids (**A**) and tetraploids (**B**). Ue—upper epidermis; Lm—leaf mesophyll; Lv—leaf vein; Le—lower epidermis; and S—stomata. Bars = 200 μm.

### 3.4. Changes in DNA Content

Histograms from the simultaneous analysis of nuclear suspensions from *A. rosea* and the internal reference standard (*Z. mays* 'B73') stained with PI displayed clear fluorescence peaks, confirming successful DNA content estimation. The mean CV values for both the reference and sample plantlets were below 5%. By utilizing *Z. mays* 'B73' as an internal standard (Figure 6A) in conjunction with diploids (Figure 6B) and tetraploids (Figure 6C), the average 2C DNA contents were found to be 2.90 and 5.62 Gb, respectively. The tetraploids exhibited a genome size that was 1.94 times larger than that of their diploid counterparts.

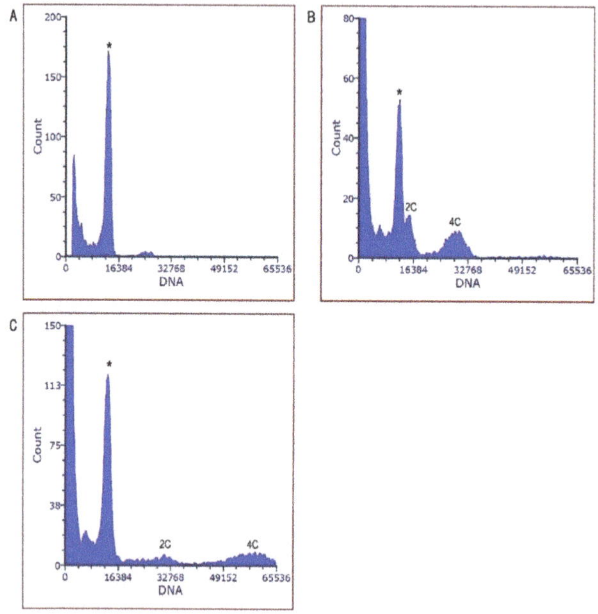


**Figure 6.** Genome size of *A. rosea* was estimated via FCM. DNA histograms of PI-stained nuclei of internal reference standard *Z. mays* 'B73', (**A**) the diploid, (**B**) and tetraploid (**C**). Peaks of the internal reference standard in the histograms of diploids and tetraploids were indicated by an asterisk. (* represents the 2C peak of *Z. mays* 'B73').

## 4. Discussion

The artificial induction of polyploids significantly increases the frequency of polyploid plants in breeding programs. An efficient in vitro mitotic chromosome doubling protocol depends on various factors, including species, explant type, antimitotic agents, exposure concentration and time, application method, and confirmation technique [19]. Protocorm-like bodies (PLBs) and protocorms are the primary explants for in vitro polyploid induction [18]. In orchids, thousands of seeds can be obtained from a single capsule, which after one or several months of in vitro germination, become numerous protocorms. Therefore, this study used the protocorms of *A. rosea* for induction. Research indicates that using protocorms for polyploid induction in orchids is widespread and highly effective. For instance, almost all reported *Dendrobium* species have used protocorms or PLBs as in vitro induction materials, with induction efficiencies reaching up to 80% [38].

The optimum concentration and duration of colchicine exposure differ among species, even within the same genus of plants. For instance, a 0.5% colchicine treatment for 10 days induced the most polyploids in *P. amabilis* Blume and *P. amboinensis* J.J.Sm. [39], while a 0.01% concentration for the same period was optimal for *P. pulcherrima* (Lindl.) J.J.Sm [40]. Moreover, the ideal colchicine concentration and exposure time for orchid polyploid induction depend significantly on the explant type and application method used. For *C. hybrid*, a 24 h soak of stem segments in a 0.05% colchicine solution resulted in a high induction rate of 28.2% [41]. However, when treating PLBs with the same concentration, it required 5 days to achieve its highest induction rate of 23.7% [42]. In our study, a colchicine concentration of 0.2% proved to be more effective in inducing polyploidy compared to concentrations of 0.05% and 0.1%. However, considering the balance between survival and induction rates, treatment with 0.05% colchicine for 5 days was also relatively effective. This protocol yielded an acceptable induction rate while maintaining a higher survival rate among the treated protocorms.

Colchicine has been shown to significantly reduce cell viability. In this experiment, our findings indicate that higher concentrations and longer exposure durations of colchicine absorbed by protocorms result in lower survival rates for both the protocorms and the resulting plantlets. This outcome is consistent with results in many plants, such as pear (*Pyrus communis* L.), mulberry (*Morus alba* L.), and perennial ryegrass (*Lolium perenne* L.) [43–45]. Low doses over extended periods are generally considered safer and more effective for inducing polyploidy without severely compromising plant health [46]. In this study, 0.05% colchicine reduced the survival rate of protocorms to below 50%, likely due to the greater sensitivity of *A. rosea* to colchicine [47]. Moreover, although low concentrations may not immediately affect cell viability, prolonged exposure can lead to cell death and reduced proliferation rates [48]. Additionally, negative effects of polyploidy can be observed during the growth stages. Manzoor et al. [49] observed that colchicine-treated gladiolus plants exhibited a range of leaf abnormalities, with the severity of these effects increasing at higher concentrations. In our study, the diploid leaves were upright and smooth, but the colchicine-treated leaves showed varying degrees of crimping and stunted growth. Specifically, leaves treated with 0.2% colchicine were semicircular and grew particularly slowly.

In polyploid induction, soaking plant tissues in colchicine tends to increase the occurrence of chimeras, which are plants with genetically different tissues. This is likely because the treatment does not uniformly affect all cells, leading to mixed ploidy levels within the same plant. Conversely, adding colchicine to solid culture media reduces chimera formation by promoting more stable polyploidy. The solid medium allows for more consistent exposure and treatment, thereby reducing variability. For instance, research on *D. wardianum* R. Warner has shown that cultivating protocorms in a culture medium with 75 μM colchicine for 30 days resulted in a 34% mutation rate and produced more stable polyploids compared to soaking. Using solid culture media helps to ensure uniformity in ploidy induction, thus reducing the formation of chimeras. [50]. Furthermore, higher concentrations of colchicine for shorter periods can cause severe malformations and reduce the production of tetraploid plants. Therefore, moderate concentrations with longer exposure periods have proven to be more effective and less toxic, reducing side effects and enhancing the overall induction rate of stable polyploids [51]. In this study, it is reasonable to increase the duration of colchicine exposure in the solid culture medium to 5–10 days.

The artificial induction of chromosome doubling in orchids, particularly through the use of colchicine, results in significant changes in stomatal and anatomical characteristics. Polyploid orchids typically exhibit larger stomata compared to their diploid counterparts. While the size of stomata increases, stomatal density often decreases in polyploids [50]. These changes in stomatal characteristics can influence photosynthetic efficiency and water regulation within the plant. Larger stomata can facilitate increased gas exchange, while reduced density might help minimize water loss under stressful conditions. Chromosome doubling leads to an overall increase in cell size, contributing to thicker leaves and stems [18]. Research indicates that colchicine treatment can lead to approximately a one-third reduction in plant height [9]. The extent of height reduction is influenced by the concentration of colchicine used, with higher concentrations exacerbating growth inhibition and reducing overall plant vigor [50]. Consequently, both plant morphology and stomatal features have been established as reliable indirect indicators for determining the ploidy level of *A. rosea*.

Moreover, endoreduplication, a phenomenon where cells have multiple sets of chromosomes, is prevalent in various orchid species [52]. In orchids, endoreduplication is linked to the rate of plant growth [53]. Tissue culture techniques that use endopolyploid cells, involving successive cycles of sectioning protocorms, have been adopted. For example, Chen observed endopolyploidy in diploid *P. aphrodite* Rchb. f. protocorms. Protocorms, measuring 3–5 mm in length after germination, were horizontally sectioned into equal parts before the expansion of the first leaf. The basal portion, with the sectioning surface facing upward, was placed into the culture medium for ex vitro regeneration. FCM determined that 34% of the PLBs in the first cycle of regeneration from the protocorms were polyploid [54]. In this study, endopolyploid cells were also detected in genotype-regenerated plantlets derived from diploid control protocorms. The ploidy level analysis showed that they were all diploids.

The genome size of artificially doubled plants does not always increase in direct multiples of the diploid genome. In our study, we used FCM with PI as the DNA fluorochrome on young leaf tissue to estimate the genome sizes of both solid diploids and colchicine-induced tetraploids. The tetraploids exhibited 1.94 times the DNA content of the diploids, with genome sizes of 2.90 and 5.62 Gb, respectively. These findings suggest that DNA loss or genome downsizing may occur after polyploid formation [55]. Furthermore, the extensive endoreduplication in orchid somatic tissues complicates genome size analysis. This challenge affects both whole and partial genome assessments, and the potential absence of 2C DNA content nuclei in the G1 phase could lead to the overestimation of genome size values. To address this, it is essential to randomly select multiple plants for analysis, with each plant being analyzed multiple times. Repeated measurements from the same plant can detect process variations and estimate differences between plants. While traditionally used leaves might yield inaccurate C-values, especially in species with gradual partial endoreduplication, young ovaries and pollinated pollen consistently show clear 2C and 1C peaks for G1 phase nuclei, making them optimal for orchid genome size studies, followed by leaf tips [56]. In this study, since the induced polyploid plants did not flower, leaf tips were used to measure genome size.

## 5. Conclusions

In this study, we report for the first time the induction of autotetraploids in *A. rosea* by treating diploid protocorm explants with colchicine in vitro. The appropriate concentration of colchicine and exposure time are crucial for increasing the rate of tetraploid induction. Based on our findings, the most effective approach, considering the balance between survival and induction rates, was treatment with 0.05% colchicine for 5 days. Compared to diploids, the tetraploids exhibited significant changes not only in morphological and anatomical characteristics but also in genome size. These tetraploid plants have the potential to serve as progenitors in the creation of novel fragrant germplasms, offering a diverse array of fragrances within the subtribe Aeridinae breeding programs.

**Author Contributions:** Conceptualization: R.J. and X.Y.; Methodology: L.W. and P.Z.; Software and Validation: L.W., Y.K. and R.J.; Investigation: L.W. and X.Z.; Formal Analysis: L.W., P.Z. and R.J.; Resources: R.J.; Data Curation: L.W. and R.J.; Writing—Original Draft Preparation: L.W.; Writing—Review and Editing: S.Y. and H.G.; Visualization: L.W. and P.Z.; Supervision: R.J. and X.Y. All authors have read and agreed to the published version of the manuscript.

**Funding:** This work was performed with the support of the National Key R&D Program of China (2023YFD1600505) and (2021YFD1200205).

**Data Availability Statement:** All data are included in the main text.

**Conflicts of Interest:** The authors declare no conflicts of interest.

## References

1. Yang, F.; Gao, J.; Li, J.; Wei, Y.; Xie, Q.; Jin, J.; Lu, C.; Zhu, W.; Wong, S.-M.; Zhu, G. The China orchid industry: Past and future perspectives. *Ornam. Plant Res.* **2024**, *4*, e002. [CrossRef]
2. Swarts, N.D.; Dixon, K.W. Terrestrial orchid conservation in the age of extinction. *Ann. Bot.* **2009**, *104*, 543–556. [CrossRef] [PubMed]
3. U.S. Fish & Wildlife Service. *Convention on International Trade in Endangered Species of Wild Fauna and Flora (CITES)*; U.S. Fish & Wildlife Service: Washington, DC, USA, 1973.
4. Pridgeon, A.M.; Cribb, P.; Chase, M.W.; Rasmussen, F.N. *Genera Orchidacearum Volume 6: Epidendroideae (Part 3)*; OUP: Oxford, UK, 2014.
5. Kishor, R.; Sha Valli Khan, P.S.; Sharma, G.J. Hybridization and in vitro culture of an orchid hybrid *Ascocenda* 'Kangla'. *Sci. Hortic.* **2006**, *108*, 66–73. [CrossRef]

6. Kocyan, A.; Vogel, E.F.D.; Conti, E.; Gravendeel, B. Molecular phylogeny of *Aerides* (Orchidaceae) based on one nuclear and two plastid markers: A step forward in understanding the evolution of the Aeridinae. *Mol. Phylogenetics Evol.* **2008**, *48*, 422–443. [CrossRef]
7. De, L.; Rao, A.; Rajeevan, P.; Pathak, P. Orchid improvement—An overview. *J. Orchid. Soc. India* **2014**, *28*, 35–45.
8. Orchid Register. Available online: https://apps.rhs.org.uk/horticulturaldatabase/orchidregister/orchidregister.asp (accessed on 3 December 2024).
9. Manzoor, A.; Ahmad, T.; Bashir, M.A.; Baig, M.M.Q.; Quresh, A.A.; Shah, M.K.N.; Hafiz, I.A. Induction and identification of colchicine induced polyploidy in 'White Prosperity'. *Folia Hortic.* **2018**, *30*, 307–319. [CrossRef]
10. Bharadwaj, D.N. Polyploidy in Crop Improvement and Evolution. In *Plant Biology and Biotechnology: Volume I: Plant Diversity, Organization, Function and Improvement*; Bahadur, B., Venkat Rajam, M., Sahijram, L., Krishnamurthy, K.V., Eds.; Springer: New Delhi, India, 2015; pp. 619–638.
11. Ramsey, J.; Schemske, D.; Ramsey, J.; Schemske, D.W. Pathways, mechanisms, and rates of polyploid formation in flowering plants. *Annu. Rev. Ecol. Syst.* **1998**, *29*, 467–501. [CrossRef]
12. Eng, W.-H.; Ho, W.-S. Polyploidization using colchicine in horticultural plants: A review. *Sci. Hortic.* **2019**, *246*, 604–617. [CrossRef]
13. Osborn, T.C.; Pires, J.C.; Birchler, J.A.; Auger, D.L.; Chen, Z.J.; Lee, H.S.; Comai, L.; Madlung, A.; Doerge, R.W.; Colot, V.; et al. Understanding mechanisms of novel gene expression in polyploids. *Trends Genet.* **2003**, *19*, 141–147. [CrossRef]
14. Silva, P.A.K.X.D.M.E.; Callegari-Jacques, S.M.; Bodanese-Zanettini, M.H. Induction and identification of polyploids in *Cattleya intermedia* Lindl. (orchidaceae) by in vitro techniques. *Cienc. Rural* **2000**, *30*, 105–111. [CrossRef]
15. Miguel, T.P.; Leonhardt, K.W. In vitro polyploid induction of orchids using oryzalin. *Sci. Hortic.* **2011**, *130*, 314–319. [CrossRef]
16. Chen, W.-H.; Chen, H.-H. Endopolyploidy in *Phalaenopsis* Orchids and Its Application in Polyploid Breeding. In *Orchid Biotechnology II*; World Scientific Publishing Company: Singapore, 2011; pp. 32–48.
17. Germana, M. Use of irradiated pollen to induce parthenogenesis and haploid production in fruit crops. In *Plant Mutation Breeding and Biotechnology*; CABI: Wallingford, UK, 2012; pp. 411–421. [CrossRef]
18. Vilcherrez-Atoche, J.A.; Iiyama, C.M.; Cardoso, J.C. Polyploidization in Orchids: From Cellular Changes to Breeding Applications. *Plants* **2022**, *11*, 469. [CrossRef] [PubMed]
19. Dhooghe, E.; Van Laere, K.; Eeckhaut, T.; Leus, L.; Van Huylenbroeck, J. Mitotic chromosome doubling of plant tissues in vitro. *Plant Cell Tissue Organ Cult. (PCTOC)* **2011**, *104*, 359–373. [CrossRef]
20. Sattler, M.C.; Carvalho, C.R.; Clarindo, W.R. The polyploidy and its key role in plant breeding. *Planta* **2016**, *243*, 281–296. [CrossRef]
21. Griesbach, R.J. Colchicine-induced polyploidy in *Phalaenopsis* orchids. *Plant Cell Tissue Organ Cult.* **1981**, *1*, 103–107. [CrossRef]
22. Tuwo, M.; Indrianto, A. Improvement of Orchid Vanda Hybrid (*Vanda limbata* Blume X *Vanda tricolor* Lindl. var. suavis) by Colchicines Treatment In Vitro. *Mod. Appl. Sci.* **2016**, *10*, 83. [CrossRef]
23. Menezes-Sá, T.S.A.; Arrigoni-Blank, M.D.F.; Santos da Costa, A.; Santos-Serejo, J.A.D.; Fitzgerald Blank, A.; Alves Soares, C.; Santos Moura, G.M. Chromosome doubling in *Cattleya tigrina* A. Rich. *Sci. Plena* **2019**, *15*, 1–10. [CrossRef]
24. Song, L.; Yang, J.; Liu, D.; Li, Z.; Wang, Y. Polyploid induction in *Cymbidium sinenthese* 'Lv mosu'× *Cymbidium hybridum* 'Shijieheping' F1 generation. *Guihaia* **2018**, *38*, 188–194. (In Chinese) [CrossRef]
25. Zhang, X.; Gao, J. In vitro tetraploid induction from multigenotype protocorms and tetraploid regeneration in *Dendrobium officinale*. *Plant Cell Tissue Organ Cult. (PCTOC)* **2020**, *141*, 289–298. [CrossRef]
26. Zaker, F.; Kolahi, H. Induction of in vitro polyploidy in ornamental flowers of orchid species (Phalaenopsis amabilis). *Iran. J. Rangel. For. Plant Breed. Genet. Res.* **2017**, *25*, 259–270. [CrossRef]
27. Dolezel, J.; Greilhuber, J.; Suda, J. Estimation of nuclear DNA content in plants using flow cytometry. *Nat. Protoc.* **2007**, *2*, 2233–2244. [CrossRef] [PubMed]
28. Mochizuki, A.L.; Sueoka, N. Genetic Studies on the Number of Plastid in Stomata. *Cytologia* **1955**, *20*, 358–366. [CrossRef]
29. Ye, Y.M.; Tong, J.; Shi, X.P.; Yuan, W.; Li, G.R. Morphological and cytological studies of diploid and colchicine-induced tetraploid lines of crape myrtle (*Lagerstroemia indica* L.). *Sci. Hortic.* **2010**, *124*, 95–101. [CrossRef]
30. Murashige, T.; Skoog, F. A Revised Medium for Rapid Growth and Bio Assays with Tobacco Tissue Cultures. *Physiol. Plant.* **1962**, *15*, 473–497. [CrossRef]
31. Lines, A.D. Value of the $K^+$ salt of carageenan as an agar substitute in routine bacteriological media. *Appl. Environ. Microbiol.* **1977**, *34*, 637–639. [CrossRef]
32. Salsabila, S.N.; Fatimah, K.; Noorhazira, S.; Halimatun, T.S.T.A.B.; Aurifullah, M.; Suhana, Z. Effect of Coconut Water and Peptone in Micropropagation of *Phalaenopsis amabilis* (L.) Blume Orchid. *IOP Conf. Ser. Earth Environ. Sci.* **2022**, *1102*, 012002. [CrossRef]
33. Zhang, X.; Gao, J. Colchicine-induced tetraploidy in *Dendrobium cariniferum* and its effect on plantlet morphology, anatomy and genome size. *Plant Cell Tissue Organ Cult. (PCTOC)* **2021**, *144*, 409–420. [CrossRef]
34. Lee, Y.-I.; Tseng, Y.F.; Lee, Y.-C.; Chung, M.-C. Chromosome constitution and nuclear DNA content of *Phalaenopsis* hybrids. *Sci. Hortic.* **2020**, *262*, 109089. [CrossRef]
35. Schnable, P.S.; Ware, D.; Fulton, R.S.; Stein, J.C.; Wei, F.; Pasternak, S.; Liang, C.; Zhang, J.; Fulton, L.; Graves, T.A.; et al. The B73 Maize Genome: Complexity, Diversity, and Dynamics. *Science* **2009**, *326*, 1112–1115. [CrossRef]

36. Galbraith, D.W.; Lambert, G.M.; Macas, J.; Dolezel, J. Analysis of nuclear DNA content and ploidy in higher plants. In *Current Protocols in Cytometry*; Wiley: Hoboken, NJ, USA, 2001; Chapter 7, Unit 7.6. [CrossRef]
37. Dolezel, J.; Bartos, J.; Voglmayr, H.; Greilhuber, J. Nuclear DNA content and genome size of trout and human. *Cytom. A* **2003**, *51*, 127–128. [CrossRef]
38. Grosso, V.; Farina, A.; Giorgi, D.; Nardi, L.; Diretto, G.; Lucretti, S. A high-throughput flow cytometry system for early screening of in vitro made polyploids in *Dendrobium hybrids*. *Plant Cell Tissue Organ Cult. (PCTOC)* **2018**, *132*, 57–70. [CrossRef]
39. Rahayu, E.M.D.; Sukma, D.; Syukur, M.; Rawati. Induksi Poliploidi *Phalaenopsis amabilis* (L.) Blume dan *Phalaenopsis amboinensis* J J. Smith dengan Kolkisin dalam Kultur In Vitro. *Indones. J. Agron.* **2016**, *43*, 219–226. [CrossRef]
40. Rungruchkanont, K.; Apisitwanich, S. Colchicine Treatment: A Method for Genetic Diversity Induction of *Doritis pulcherrima* Lindl. Orchid of Thailand. In Proceedings of the 2nd International Orchid Symposium, Bangkok, Thailand, 19–21 February 2014, pp. 175–180.
41. Bi-Xia, J.; Dawei, C.; Chenchen, Z.; Di, M.; Wen-jing, H.; Yu, W. High Efficient Polyploid Induction of *Cymbidium hybridium*. *Bull. Bot. Res.* **2011**, *31*, 558. [CrossRef]
42. MuGui, W.; Ruizhen, Z.; Li, X.; Yanghui, L.; FeiYan, Z.; BaoGui, D.; Zhisheng, Z. In vitro induction and its identification of tetraploid *Cymbidium hybridium*. *Acta Bot. Boreali-Occident. Sin.* **2010**, *30*, 56–62. (In Chinese)
43. Chakraborti, S.P.; Vijayan, K.; Roy, B.N.; Qadri, S.M.H. In vitro induction of tetraploidy in mulberry (*Morus alba* L.). *Plant Cell Rep.* **1998**, *17*, 799–803. [CrossRef]
44. Nair, R.M.; Cooperative, R. Developing tetraploid perennial ryegrass (*Lolium perenne* L.) populations. *New Zealand J. Agric. Res.* **2004**, *47*, 45–49. [CrossRef]
45. Sun, Q.; Sun, H.; Li, L.; Bell, R.L. In vitro colchicine-induced polyploid plantlet production and regeneration from leaf explants of the diploid pear (*Pyrus communis* L.) cultivar, 'Fertility'. *J. Hortic. Sci. Biotechnol.* **2009**, *84*, 548–552. [CrossRef]
46. Samatadze, T.E.; Yurkevich, O.Y.; Khazieva, F.M.; Basalaeva, I.V.; Konyaeva, E.A.; Burova, A.E.; Zoshchuk, S.A.; Morozov, A.I.; Amosova, A.V.; Muravenko, O.V. Agro-Morphological and Cytogenetic Characterization of Colchicine-Induced Tetraploid Plants of *Polemonium caeruleum* L. (Polemoniaceae). *Plants* **2022**, *11*, 2585. [CrossRef]
47. Sinha, S.K. In Vitro Studies on the Effect of Colchicine Treatment and Gamma-Irradiation in *Renades* 'Arunodaya' Hybrid and Its Parent Species *Aerides rosea* Loddiges ex Paxt. (*A. williamsii* Warn.) and *Renanthera imschootiana* Rolfe. Ph.D. Thesis, Gauhati University, Guwahati, India, 2000.
48. Wang, D.; Xie, Y.; Yan, M.; Pan, Q.; Liang, Y.; Sun, X. Colchicine causes prenatal cell toxicity and increases tetraploid risk. *BMC Pharmacol. Toxicol.* **2019**, *20*, 66. [CrossRef]
49. Manzoor, A. *Induction and Identification of Polyploidy in Gladiolus Grandiflorus Through Chemical Treatment*; Arid Agriculture University: Rawalpindi, Pakistan, 2016.
50. Wang, F.; Zhuo, X.; Arslan, M.; Ercisli, S.; Chen, J.; Liu, Z.; Lan, S.; Peng, D. In Vitro Induction of Polyploidy by Colchicine in the Protocorm of the Orchid *Dendrobium wardianum* Warner. *HortScience* **2023**, *58*, 1368–1375. [CrossRef]
51. Manzoor, A.; Ahmad, T.; Bashir, M.A.; Hafiz, I.A.; Silvestri, C. Studies on Colchicine Induced Chromosome Doubling for Enhancement of Quality Traits in Ornamental Plants. *Plants* **2019**, *8*, 194. [CrossRef] [PubMed]
52. Larkins, B.A.; Dilkes, B.P.; Dante, R.A.; Coelho, C.M.; Woo, Y.M.; Liu, Y. Investigating the hows and whys of DNA endoreduplication. *J. Exp. Bot.* **2001**, *52*, 183–192. [CrossRef] [PubMed]
53. Brown, S.C.; Bourge, M.; Maunoury, N.; Wong, M.; Wolfe Bianchi, M.; Lepers-Andrzejewski, S.; Besse, P.; Siljak-Yakovlev, S.; Dron, M.; Satiat-Jeunemaître, B. DNA Remodeling by Strict Partial Endoreplication in Orchids, an Original Process in the Plant Kingdom. *Genome Biol. Evol.* **2017**, *9*, 1051–1071. [CrossRef] [PubMed]
54. Chen, W.-H.; Tang, C.-Y.; Lin, T.-Y.; Weng, Y.-C.; Kao, Y.-L. Changes in the endopolyploidy pattern of different tissues in diploid and tetraploid *Phalaenopsis aphrodite* subsp. *formosana* (Orchidaceae). *Plant Sci.* **2011**, *181*, 31–38. [CrossRef] [PubMed]
55. Leitch, I.J.; Bennett, M.D. Genome downsizing in polyploid plants. *Biol. J. Linn. Soc.* **2004**, *82*, 651–663. [CrossRef]
56. Trávníček, P.; Ponert, J.; Urfus, T.; Jersáková, J.; Vrána, J.; Hřibová, E.; Doležel, J.; Suda, J. Challenges of flow-cytometric estimation of nuclear genome size in orchids, a plant group with both whole-genome and progressively partial endoreplication. *Cytom. A* **2015**, *87*, 958–966. [CrossRef]

plants

MDPI

Article

# Cryopreservation Protocols and the Associated Ultrastructural Changes in Dormant Buds of *Vitis amurensis*

Dan Sun, Peijin Ni, Jian Liu, Zhenxing Wang, Guangli Shi, Meng Li, Xuanhe Zhang and Jun Ai *

College of Horticulture, Jilin Agricultural University, Changchun 130118, China; sundan@jlau.edu.cn (D.S.); zhenxinghd@aliyun.com (Z.W.)
* Correspondence: aijun1005@163.com

**Abstract:** There is an urgent need for the cryopreservation of dormant buds to conserve the genetic resources of woody plants, particularly fruit trees, as this method is less time-consuming and relatively inexpensive. In the present study, three different cryopreservation protocols were tested on dormant buds from three varieties of *Vitis amurensis* Rupr. The explants were collected between November 2017 and March 2018. Twig segments harvested from field-grown plants, each containing one dormant bud, were desiccated in a low-temperature test chamber at −5 °C. The viability of the buds was highest (45%) after 28–30 days of desiccation, when the moisture content was approximately 25–30%. Cryopreservation using the CP3 protocol (which involves decreasing the temperature at a rate of 0.1 °C/min to −30 °C and holding this temperature for 24 h, followed by a 0.5 °C/min decline to −80 °C, a 1 °C/min decline to −180 °C, and finally reaching −196 °C in a CryoMed controlled rate freezer) significantly enhanced the viability (66.67%) when the samples were packed in aluminum-foil bags. Additionally, immersing the twigs in ice-cold (4 °C) water for 24 h in a refrigerator during thawing proved to be more conducive to viability. The dormant buds of all three *V. amurensis* varieties collected in January exhibited the highest viability after cryopreservation, followed by those collected in February and December. In contrast, the dormant buds collected in November and March showed the lowest viability after cryopreservation. The average viability of twigs of 'Shuanghong', 'Zuoshanyi', and 'Shuangfeng' collected between 2019 and 2021 all exceeded 60%. After the cryopreservation process, the outer multilayered cells in the buds were completely damaged; however, the inner cells exhibited moderate damage and were able to resume growth after thawing. Therefore, based on graft viability and histological observations, the dormant bud cryopreservation protocols tested in this study could be applicable to these three *V. amurensis* varieties.

**Keywords:** conservation of resources; dormant buds; viability; *Vitis amurensis*; structural changes

## 1. Introduction

The conservation and availability of a wide variety of fruit germplasm including varieties and wild relatives require a strong genetic resource pool. Traditionally, field gene pools serve as ex situ resources for problematic materials. However, several limitations that hinder their utility pose significant threats to germplasm security [1]. Genetic resources are vulnerable to pests, diseases, natural disasters such as drought, human error, and vandalism. Furthermore, germplasm exchange under these conditions carries substantial risks due to the potential transfer of diseases through vegetative propagation [1,2]. Current in vitro gene pools offer a viable option for the short-term and medium-term preservation of various crops [3] including temperate fruit trees and horticultural species [4–7]. Nevertheless, these in vitro gene pools are not ideal for long-term preservation, as this method can be costly and may result in resource loss due to contamination or technical challenges during regrowth intervals.

Cryopreservation is a valuable tool for the long-term storage of clonally propagated plants and seeds [8,9]. Significant advancements in plant cryopreservation have occurred since the initial experiments conducted from the 1970s to the 1990s, leading to the development of various cryopreservation techniques since the 2000s. Numerous studies on germplasm cryopreservation have been published for a wide range of plant species including herbs, vines, shrubs, and trees across diverse growth regions, from tropical and subtropical to temperate climates [8,10]. Propagules such as shoot tips and dormant buds are candidates for clonal cryopreservation, in which the traits of interest from the parent plant are retained [11,12]. The cryopreservation of dormant buds (DBs) is less time-consuming than that of the shoot tips, as DBs do not require aseptic culture or additional time for excision, increasing the cryopreservation capacity [13]. In a pioneering study, Sakai initiated the cryopreservation dormant buds on *Salix* and *Populus*, and then extended this to other species such as *Vitis* [14–17]. Since then, dormant bud cryopreservation has been used for other species such as *Betula pendula* [18], *Prunus cerasus* [19], blueberry [20], *Malus* [21], *Ulmus* [22], *Fraxinus* [23], Pyrus [24], *Pinus sylvestris* [25], and other tree species. Although there may have been successful scientific reports, the dormant bud approach has not been widely adopted for the preservation of crop germplasm. One possible reason for this is that the method is primarily utilized for shrubs and trees that can survive dormancy; however, the DB of crops may not yield satisfactory post-cryopreservation viability.

*Vitis amurensis* Rupr. (Vitaceae), a nature-state vine, is widely distributed from the northeastern regions of China to northern Korea. It is considered as one of the most desirable wine grape varieties in China [26]. Modern pharmacological studies indicate that *V. amurensis* exhibits anti-inflammatory [27], antimicrobial [28], antioxidant [29,30], and anticancer properties [31,32]. *V. amurensis* vines are generally dioecious, although hermaphroditic vines have also been occasionally reported. This plant species demonstrates high resistance to cold temperatures [33], allowing the viability of branches and roots at −40 °C and −16 °C, respectively [34]. Consequently, *V. amurensis* is regarded as one of the most valuable germplasm resources in grape breeding. The cryopreservation of *Vitis* shoot tips and dormant buds has been reported in some species, but most studies have been carried out on shoot tips, and only one study has been published on dormant buds [35–38]. Limited results were obtained when *Vitis* dormant buds were cryopreserved. To the best of our knowledge, there are currently no reports on the cryopreservation of *V. amurensis* germplasm resources. The cryopreservation of DB is more cost-effective, and the recovery process is simpler, utilizing direct grafting or tissue culture [39,40]. In this study, we investigated the optimal conditions for the cryopreservation of DB of *V. amurensis*. We tested various cryopreservation protocols on three varieties of *V. amurensis* and examined the ultrastructural changes during the successive steps of desiccation prior to cryopreservation as well as after thawing. The findings of this study provide effective guidelines, protocols, and methods for the conservation and sustainable utilization of *V. amurensis*.

## 2. Materials and Methods

### 2.1. Plant Material

A total of 1800 one-year-old dormant twigs from 30 individual plants of *V. amurensis* 'Shuanghong', 'Zuoshanyi', and 'Shuangfeng' (planted in 2010) were collected between November (autumn) 2017 and March (spring) 2018, and in January (winter) 2019, 2020, and 2021 at the orchard of the Jilin Agricultural University (latitude 43°86′; longitude 125°28′), Changchun, China. The lowest and average temperatures are shown in Figures 1 and 2. Twigs were cut into 70 mm segments, and each segment contained one dormant bud, which was 1 cm away from the top. The basal and uppermost part of the shoots were not used. The top diameter of the twigs of between 6 and 7 mm was used.

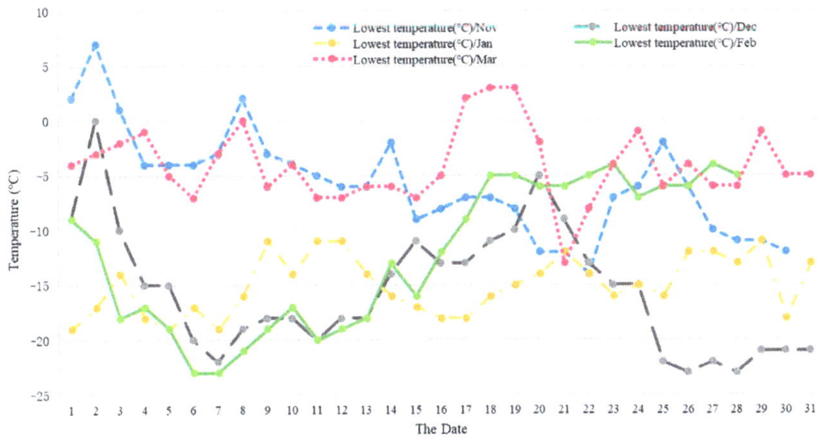


**Figure 1.** The lowest temperature profile at the orchard of the Jilin Agricultural University (latitude 43°86′; longitude 125°28′), Changchun, China from November 2017 to March 2018.

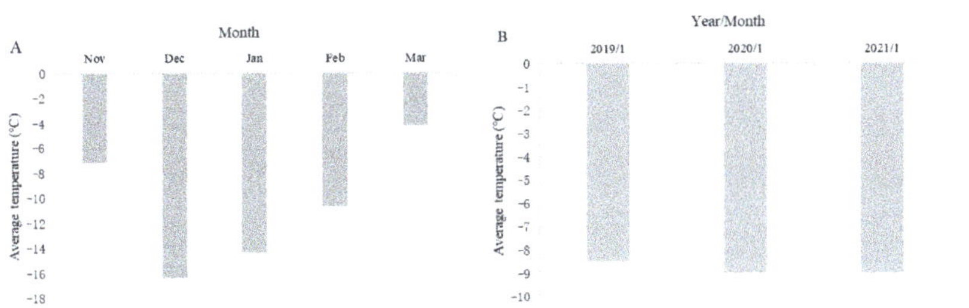


**Figure 2.** The average temperature at the orchard of the Jilin Agricultural University (latitude 43°86′; longitude 125°28′), Changchun, China. (**A**) The average temperature from November 2017 to March 2018. (**B**) The average temperature of January in 2019, 2020, and 2021.

The standard cryopreservation procedure was as follows (Figure 3). The twigs were desiccated in a low-temperature test chamber at −5 °C for 28 days to the moisture content of approximately 25–30%. Three twigs were packed in aluminum-foil bags and then transferred to a CryoMed controlled rate freezer for cooling down at a rate of 0.1 °C/min to −30 °C, maintaining this temperature for 24 h, followed by a decline of 0.5 °C/min to −80 °C, then a decline of 1 °C/min to −180 °C, and finally dropped into LN. The thawing procedure consisted of immersing the twigs in ice-cold (4 °C) water for 24 h. Twig segments were then grafted onto the rootstock Beta "Zaohong" (1~2-year-old) in a greenhouse maintained at 25 ± 3 °C and 75–95% relative humidity. Rootstocks were selected for grafting based on their thickness, which should match the thickness of the scion. The rootstock was cut flat approximately 5 cm from the rhizome, and the bud eye was removed. A rootstock was split about 1 cm from the center of the stem. The lower part of the scion was split into two opposing inclined planes, each about 1 cm in length. The cut end of the scion was inserted into the incision made in the rootstock, ensuring proper alignment. Finally, the joined sections were secured with thin plastic strips. The buds' viability was evaluated as the number of buds that sprouted five weeks after grafting.

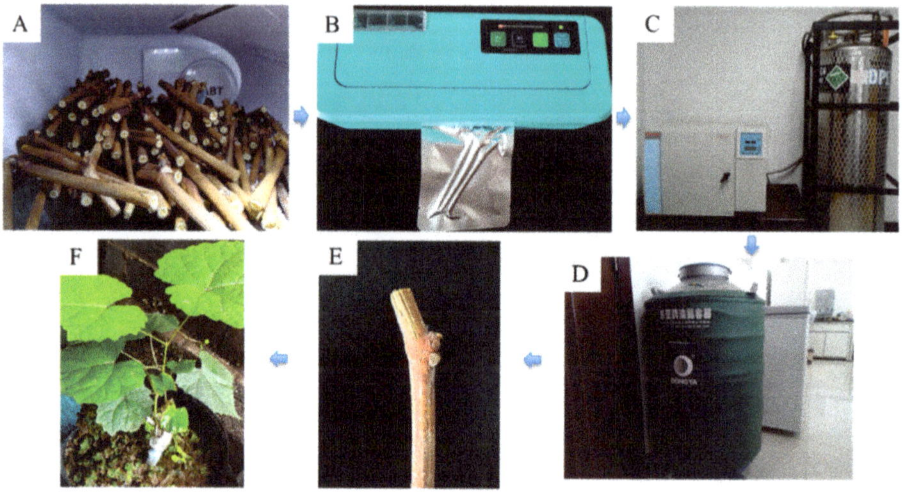


**Figure 3.** Cryopreservation process of the dormant twigs of *V. amurensis*. (**A**) Twig desiccation at 5 °C in a refrigerator; (**B**) desiccated twigs were vacuumed packed with aluminum-foil bag; (**C**) desiccated twigs frozen in a CryoMed controlled rate freezer; (**D**) desiccated twigs cryopreserved in LN; (**E**) cryopreserved twigs after thawing; (**F**) twigs that survived after grafting for two months.

### 2.2. The Effect of the Desiccation Procedure

Twigs of 'Shuanghong' were desiccated in a low-temperature test chamber (DW-100, Yashilin) at −5 °C for 7, 14, 20, 25, 28, 30, or 32 d. Twigs without desiccation were used as the control. Water loss during desiccation and bud viability were determined at each time point. To construct a desiccation curve, the twigs were weighed immediately after desiccation and after drying at 60 °C until reaching a constant weight. The moisture content (MC) was calculated as follows:

$$MC = (FW - DW)/FW * 100\%$$

Desiccated twigs were packed with aluminum-foil bags and immersed in liquid nitrogen (LN) for 28 d. To thaw, the twigs were removed from the aluminum-foil bags and immersed in ice-cold water (4 °C) for 24 h. The viability of the buds was determined as indicated in Section 2.1. Each treatment contained 30 twigs, and all experiments and analyses were performed in triplicate.

### 2.3. The Effect of Packing and Cryopreservation Procedure

Desiccation was performed as described above, and the amount of water loss and corresponding bud viability were determined at 28 d. Subsequently, the desiccated twigs of 'Shuanghong' were either sealed in cryovials (Nunc™, Thermo Fisher Scientific®, Waltham, MA, USA) or vacuum sealed in aluminum-foil bags; desiccated twigs subjected to no packing were used as controls. The twig samples were then treated with a different cryopreservation procedure—CP1, CP2, or CP3. For CP1, twigs were immersed in LN directly; for CP2, twigs were exposed to a decreasing temperature gradient of −10 °C for 6 h, −20 °C for 6 h, and −30 °C for 6 h in a low-temperature test chamber (DW-100, Yashilin, Beijing, China), −80°C in an ultra-low temperature freezer (DW-861626, Haier, Qingdao, China) for 6 hm and transferred to LN thereafter; for CP3, twigs were cooled down at a rate of 0.1 °C/min to −30 °C for 24 h, followed by a 0.5 °C/min decline to −80 °C and a 1 °C/min decline to −180 °C in a CryoMed controlled rate freezer (Forma7451TF, Thermo Forma, Waltham, MA, USA). Then, the twigs were submerged in LN for 28 d.

To thaw, the twigs were removed from the aluminum-foil bags and immersed in ice-cold water (4 °C) for 24 h. The viability of the buds was determined as indicated in Section 2.1.

### 2.4. The Effect of Thawing and Regrowth Procedures

The moisture content of the twigs was measured at approximately 28–30%. Desiccated twigs were packed in aluminum-foil bags and subsequently cryopreserved using the CP3 procedure described in Section 2.3. These twigs of the 'Shuanghong' variety were stored in LN for a duration of 28 d, as previously described. Following this storage period, the twigs were thawed using one of three methods: rapid thawing (10 min immersion in a water bath at 40 °C), room temperature thawing (30 min immersion in water at 25 °C), or slow thawing (24 h immersion in water at 4 °C in a refrigerator). During the slow thawing process, the water was replaced every hour for the initial 6 h. The thawed twigs were subsequently utilized for grafting purposes. The viability of the buds was determined as indicated in Section 2.1.

### 2.5. The Effect of Varieties and Harvest Time

Cryopreservation was conducted using buds from various cultivars, specifically 'Shuanghong', 'Zuoshanyi', and 'Shuangfeng', which were collected during the months of November, December, January, February, and March. The experimental procedures commenced in November and extended through to March of the following year. Desiccation of the twigs was achieved by subjecting them to a low temperature test at −5 °C. The desiccated twigs were subsequently vacuum packed in aluminum-foil bags and frozen according to the CP3 protocol. Following this, the twigs were stored in LN for a duration of 28 d. Thawing was performed slowly by immersing the twigs in water for 24 h at a temperature of 4 °C. The viability of the buds was determined as indicated in Section 2.1.

### 2.6. Sample Preparation for Transmission Electron Microscopy (TEM)

To assess the effects of cryopreservation on the cell and organelle structure, the ultrastructure of cells in the apical meristem was observed under a transmission electron microscope. An assessment was made for the injury degree and seriousness in the duration of the desiccation–freezing–thawing cycle. Frozen buds (1–3 $mm^3$) stored in LN for 28 d were sampled under sterile conditions and fixed in 4% glutaraldehyde by vacuum infiltration. The fixed buds were incubated at 5 °C overnight and then rinsed with 0.1 mol/L phosphate buffer (pH 6.8) five times, once each for 5, 10, 15, 20, and 30 min. Subsequently, the buds were fixed in 1% osmium acid at 4 °C for 2 h and rinsed 4–5 times with 0.1 mol/L phosphate buffer (pH 6.8) for 10 min each. Then, the buds were subjected to gradient ethanol dehydration as follows: 30% for 15 min, 50% 15 min, 70% ethanol for 15 min, 80% for 20 min, 90% for 20 min, and 100% for 30 min. Finally, the buds were dehydrated with 100% acetone twice, each for 30 min. The dehydrated buds were penetrated with acetone:embedding agent (Spon812) (3:1) for 2–3 h, acetone:embedding agent (1:1) for 4–6 h, acetone:embedding agent (3:1) for 12 h, and twice with pure embedding agent for 24 h each. The embedding buds were oven dried at 30 °C for 12–24 h and then at 60 °C for 24–48 h. After sample preparation, the embedding blocks were restored and positioned with half-thin parts. Samples were double stained with uranium acetate and lead citrate, and the structural changes in the cells and organelles were observed and photographed by a JEM—1400 TEM.

### 2.7. Experimental Design and Data Analyses

Each treatment contained 30 twigs, and all experiments and analyses were performed in triplicate. An analysis of variance and comparison of means were carried out using SPSS 22.0 (SPSS Inc. Chicago, IL, USA), and significant differences were determined at $p < 0.05$ using Duncan's multiple range test.

## 3. Results

### *3.1. Effects of Drying Duration and Moisture Content on the Viability of V. amurensis Buds*

The moisture content of plant tissues is a critical factor that influences their viability following cryopreservation. Figure 4 demonstrates that after 14 days of dehydration at −5 °C, the moisture content of the twigs exceeded 60%, resulting in a viability rate of 0% for the DBs post-cryopreservation. After 20 days of dehydration, the moisture content of the twigs decreased to 50%, and the viability of the DBs after cryopreservation was recorded at 13.33%. As the moisture content decreased, the viability of the buds post-cryopreservation increased. The highest bud viability, at 45%, was observed within a moisture content range of 25–30%, with an average moisture content of 28%. However, with further reductions in moisture content, the viability of the DBs also declined.

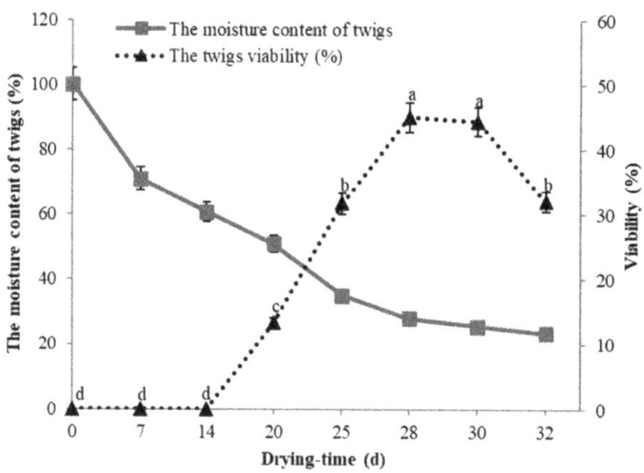


**Figure 4.** Effects of different drying times on the moisture content in twigs of 'Shuanghong' collected in January 2018 and the viability of the DBs after the cryopreservation of *V. amurensis*. Note: Mean values in each point with the same letter were not significantly different at $p < 0.05$ by the Duncan's multiple range test.

### *3.2. Effects of Packing and Cryopreservation Protocols on the Viability of V. amurensis Buds*

The procedures employed for both packing and cryopreservation had a significant impact on the viability of the DBs (Table 1). DBs that were unpacked, those stored in cryovials, and those enclosed in aluminum-foil bags demonstrated viability rates of 45.56%, 53.33%, and 66.67%, respectively, when subjected to the CP3 cryopreservation procedure. Additionally, when comparing twigs that were unpacked and those stored in cryovials, it was observed that the twigs vacuum sealed in the aluminum-foil bags exhibited superior bud viability (66.67%) across all three cryopreservation protocols (CP1, CP2, and CP3). In summary, twigs that were vacuum sealed in aluminum-foil bags and cryopreserved using the CP3 protocol exhibited the highest viability rates. Furthermore, twigs stored in cryovials and vacuum sealed in aluminum-foil bags demonstrated enhanced bud viability with the CP2 procedure in comparison to CP1; conversely, twigs that were unpacked showed greater bud viability with the CP1 procedure relative to CP2.

**Table 1.** Effects of different packing modes and cryopreservation procedure on the viability of the dormant buds of 'Shuanghong' collected in January 2018 after cryopreservation.

| Packing Mode | Moisture Content in Twigs (%) | Cryopreservation Procedure | Average Viability of Cryopreserved Buds (%) ± Standard Errors (SD) |
|---|---|---|---|
| No packing | 28~30 | CP1 | 30.00 ± 2 de |
| | 28~30 | CP2 | 25.56 ± 6 e |
| | 28~30 | CP3 | 45.56 ± 8 c |
| Cryovial | 28~30 | CP1 | 33.33 ± 2 d |
| | 28~30 | CP2 | 36.67 ± 2 d |
| | 28~30 | CP3 | 53.33 ± 4 b |
| Aluminum foil bag | 28~30 | CP1 | 42.22 ± 4 cd |
| | 28~30 | CP2 | 48.89 ± 5 c |
| | 28~30 | CP3 | 66.67 ± 4 a |

Note: Mean values in each column with the same letter were not significantly different at $p < 0.05$ by the Duncan's multiple range test.

CP1: Samples were immersed in LN directly; CP2: Samples were subjected to a decreasing temperature gradient: −10 °C for 6 h, −20 °C for 6 h, and −30 °C for 6 h in a low temperature test chamber, followed by −80 °C in an ultra-low temperature freezer for 6 h, and finally immersed in LN; CP3: Samples were processed using a CryoMed controlled rate freezer, which decreased the temperature at a rate of 0.1 °C per min to −30 °C, maintaining this temperature for 24 h. This was followed by a temperature decline of 0.5 °C per min to −80 °C, then a decline of 1 °C per min to −180 °C, before being transferred to LN.

### 3.3. *Effect of Thawing Procedures on the Viability of V. amurensis Buds*

The thawing method has a significant effect on the recovery of DBs. Specifically, DBs that underwent a 24 h slow thaw at 4 °C demonstrated a markedly higher viability rate of 66.67%. In contrast, the viability rates for fast thawing at 40 °C and at room temperature were only 20% and 13.33%, respectively (Figure 5). To improve water absorption and reduce the dissolution of pectin from the damaged cell walls resulting from cryopreservation, we implemented a protocol to change the water every hour during the initial 6 h of the slow thawing process.

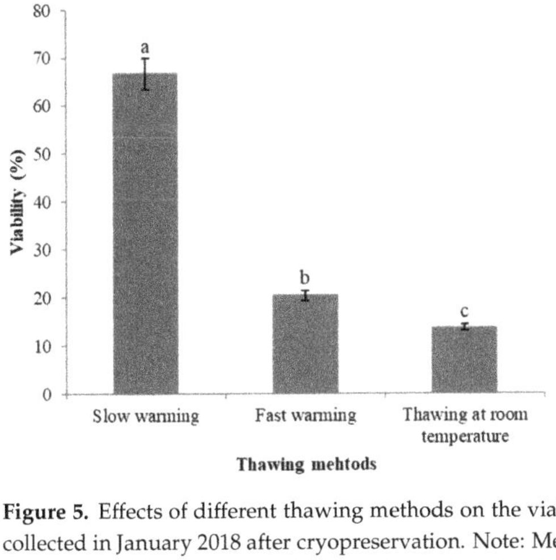


**Figure 5.** Effects of different thawing methods on the viability of dormant buds of 'Shuanghong' collected in January 2018 after cryopreservation. Note: Mean values with the same letter were not significantly different at $p < 0.05$ by the Duncan's multiple range test.

### 3.4. Varieties and Harvest Time on the Viability of *V. amurensis* Buds

DBs harvested at different times exhibited significant variation in regeneration rates, ranging from 24.44% to 66.67% after cryopreservation. Buds collected from various genotypes at similar temperatures demonstrated considerable differences in their freezing tolerance. Notably, DB from all three varieties collected in December, January, and February displayed the highest viability following cryopreservation: 'Shuanghong' (55.56%, 66.67%, and 61.11%, respectively), 'Zuoshanyi' (55.56%, 60.00%, and 58.89%, respectively), and 'Shuangfeng' (56.67%, 65.56%, and 62.22%, respectively) (Table 2). We hypothesize that the DBs on twigs harvested in November were insufficiently dormant to endure cryopreservation, while those collected in mid-March had completed their dormant phase and therefore could not survive the ultra-low temperatures of LN. The average temperatures at which the twigs were harvested showed no correlation with the viability of the cryopreserved DBs. Although the viability of the cryopreserved DBs was higher in January, the average temperature in January was actually higher than that in December. Consequently, it is advised that twigs are harvested only after sufficient cold hardiness has developed. The viability of *V. amurensis* DBs after cryopreservation between 2019 and 2021 was also examined, and the average viability of all exceeded 60% (Table 3).

**Table 2.** Effects of genotype and harvest time on the viability of *V. amurensis* dormant buds after cryopreservation.

| Variety | Harvest Date (mm/dd/yy) | Total Moisture Content (%) | Total Moisture Content After 28 Days Desiccation (%) | Average Viability of CK (%) | Average Viability After Cryopreservation (%) |
|---|---|---|---|---|---|
| 'Shuanghong' | 11/10/17 | 42.16 | 29.49 | 92.22 ± 4 b | 37.78 ± 4 d |
| | 12/10/17 | 40.33 | 28.78 | 97.78 ± 2 a | 55.56 ± 8 c |
| | 01/08/18 | 41.06 | 27.78 | 83.33 ± 8 c | 66.67 ± 1 2 a |
| | 02/01/18 | 45.10 | 28.03 | 87.78 ± 3 c | 61.11 ± 6 b |
| | 03/03/18 | 51.60 | 29.00 | 87.78 ± 4 c | 36.67 ± 2 d |
| 'Zuoshanyi' | 11/10/17 | 44.63 | 27.86 | 88.89 ± 4 b | 26.67 ± 2 c |
| | 12/10/17 | 44.47 | 28.30 | 93.33 ± 5 a | 55.56 ± 9 a |
| | 01/08/18 | 43.10 | 28.04 | 84.44 ± 1 2 b | 60.00 ± 10 a |
| | 02/01/18 | 43.25 | 27.98 | 92.22 ± 4 a | 58.89 ± 3 a |
| | 03/03/18 | 48.03 | 27.56 | 83.33 ± 5 b | 37.78 ± 4 b |
| 'Shuangfeng' | 11/10/17 | 46.61 | 28.74 | 87.78 ± 7 b | 24.44 ± 4 c |
| | 12/10/17 | 48.18 | 29.01 | 82.22 ± 5 b | 56.67 ± 5 b |
| | 01/08/18 | 46.61 | 28.36 | 88.89 ± 11 b | 65.56 ± 7 a |
| | 02/01/18 | 48.18 | 28.03 | 90.00 ± 5 a | 62.22 ± 6 b |
| | 03/03/18 | 52.74 | 27.98 | 88.89 ± 4 b | 32.23 ± 4 c |

Note: CK represents the twigs before cryopreservation. Mean values in each column with the same letter were not significantly different at $p < 0.05$ by the Duncan's multiple range test.

**Table 3.** The viability of *V. amurensis* DBs after cryopreservation between 2019 and 2021.

| Variety | Moisture Content in Twigs (%) | Average Viability in 2019 (%) | Average Viability in 2020 (%) | Average Viability in 2021 (%) |
|---|---|---|---|---|
| 'Shuanghong' | 28–30% | 68.89 ± 6 a | 66.33 ± 9 a | 70.00 ± 5 a |
| 'Zuoshanyi' | 28–30% | 62.11 ± 2 b | 64.67 ± 8 b | 63.11 ± 5 c |
| 'Shuangfeng' | 28–30% | 66.67 ± 2 a | 66.12 ± 5 a | 67.44 ± 3 b |

Note: Mean values in each column with the same letter were not significantly different at $p < 0.05$ by the Duncan's multiple range test.

### 3.5. Effect of Cryopreservation on Cell Ultrastructure

Transmission electron microscopy (TEM) was utilized to investigate the effects of desiccation and cryopreservation on meristematic cells in axillary buds. The non-cryopreserved (control) DBs exhibited large, prominent nuclei, numerous clear starch granules within plastids, and an abundance of mitochondria. These buds also displayed an intact nuclear membrane, a thick cell wall, and a complete nuclear skeleton (Figure 6A,B). Furthermore, the cell wall and cell membrane were in close proximity, with the cell membrane appearing smooth, orderly, and maintaining a distinct bilayer structure in the control buds. Dehydration removes free water from cells, thereby enhancing the stability of the cell membrane and other cellular structures. Consistently, dehydration resulted in an increase in cytoplasmic thickness and the coloration of the nuclei and nucleoli, a reduction in the abundance of large vacuoles, improved visibility of the mitochondria, and a slight separation between the cell wall and cell membrane. Nevertheless, the overall structure remained intact following dehydration (Figure 6C,D).

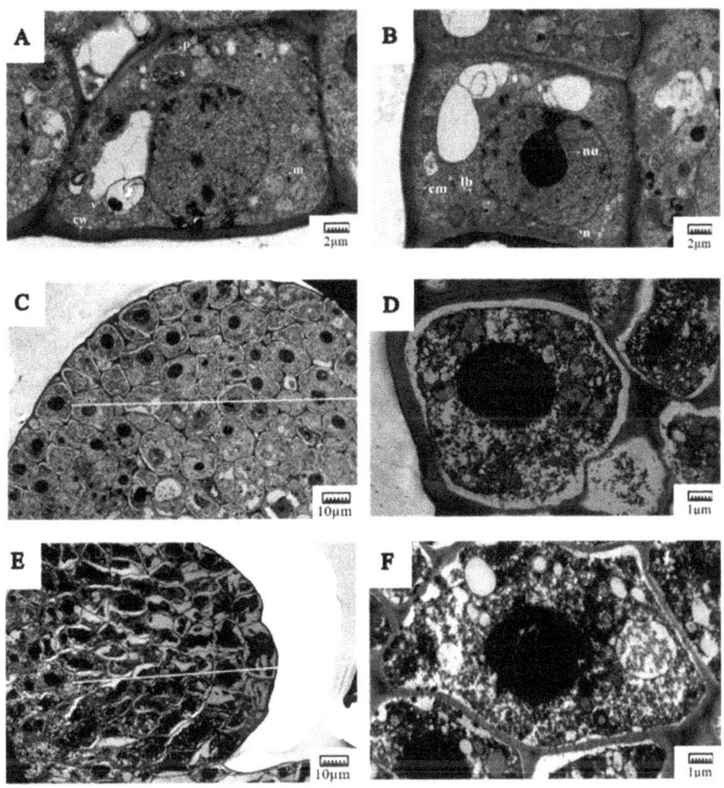


**Figure 6.** Ultrastructural observations of meristem cells in *V. amurensis* buds. (**A**,**B**) The buds before desiccation; (**C**,**D**) the buds desiccated for 28 d; (**E**,**F**) the buds after cryopreservation. Abbreviations: cm: cell membrane; cw: cell wall; lb: lipid body; m: mitochondrion; n: nucleus; nu: nucleolus; p: proplastid; s: starch; v: vacuole.

After cryopreservation, the outer six multilayered cells sustained irreparable damage, resulting in significant alterations to their structure (Figure 6E,F). In the cryopreserved samples, the gap between the cell wall and the cell membrane increased, causing the cytoplasm and nucleus to be compressed together. The plasmodesmata were disrupted, and the organelles appeared indistinct. Some cells exhibited large vacuole-like cavities filled with fibrous substances, likely representing debris formed by the disintegration of certain organelles due to drastic temperature changes, which led to lethal damage to the cells. Furthermore, the plasma membranes were compromised to varying extents, leading to an uneven surface or complete rupture. Starch, which is associated with protection against freezing injury, was not detected in the plastids. Despite some structural changes and slight plasmolysis, the endothelial cells remained intact and contained dark-colored nuclei. The endothelial cells demonstrated the capacity to resume growth following thawing, as the damage incurred during cryopreservation was assessed to be moderate. The following experiments also demonstrated that the DBs after cryopreservation had the ability to sprout after grafting (Figure 6).

## 4. Discussion

Cryopreservation protocols have been described with various propagules for grapevines such as pollen [41,42], seeds [43], somatic embryos [44–48], and shoot tips [35–37]. The method of cryopreservation for dormant buds is gaining significant interest for the preservation of genetic resources in woody plants because this method is easy to operate, and the technical requirements are not particularly high. Buds intended for cryopreservation can be collected from either field-grown or greenhouse plants of the tested genotypes. Furthermore, this technique has emerged as a viable option for cryopreservation that does not require in vitro laboratory support. In 1960, Sakai demonstrated the viability of dormant winter twigs from *Salix* and *Populus* under low-temperature conditions by slowly freezing the twigs at −30 °C before immersing them in LN [16]. Among the cryopreservation protocols developed for *Populus*, the cryostorage of dormant in vivo buds is the most suitable technique for ensuring a backup resource for field collections [49–52]. An efficient cryopreservation protocol for *V. amurensis* DBs was developed for the first time in this study.

The desiccation of twigs prior to cryopreservation requires considerable effort and time, and it is a widely used technique in many dormant bud cryopreservation protocols. Desiccation reduces the moisture content of DBs, which may decrease the cryodamage and enhance bud recovery. Dehydrating samples to a moisture content of 25–35% of their fresh weight before storage in LN has been shown to improve recovery success post-cryopreservation, depending on the species, compared to samples that were not dehydrated [53]. Vogiatzi et al. indicated that the key volume reduction in bulk water varied among apple varieties and that slow cooling under winter conditions to 30% had a significant effect on their survival [54]. According to Volk et al., *Malus* buds exhibited high viability after 10 years of storage in LN vapor, with some accessions demonstrating better viability when desiccated prior to slow cooling compared to those that were not desiccated [53]. During the cold acclimation process in winter, all apple varieties exhibited resistance to moderate dehydration stress. Höfer concluded that the initial moisture content and duration of dehydration had no impact on viability following cryopreservation, as all materials were desiccated to 30% moisture before exposure to LN [55]. However, opinions vary regarding the necessity of dehydration prior to cryopreservation. Some fruit trees do not require pre-freezing desiccation [56]. Esensee et al. reported that buds desiccated to 18% moisture content resulted in survival for certain 'Valiant' grape buds [38]. For instance, Rantala et al. suggested that dormant blackcurrant buds should retain their natural moisture content to enhance the cryopreservation process [57]. In the present study, the moisture content of dormant *V. amurensis* buds was approximately 25–30%, which was beneficial for bud viability after cryopreservation. Excessive moisture may lead to the formation of large ice crystals that can damage cellular tissue, which is detrimental to the cryopreservation of *V. amurensis* buds.

Another important parameter is the cryopreservation protocol, which can ensure high recovery rates following cryopreservation. In this study, we investigated the twigs of *V. amurensis* using three types of packaging and three cryopreservation procedures. The results indicated that twigs packed in vacuum-sealed aluminum-foil bags exhibited greater viability across all three cryopreservation methods compared to the other two packaging types. Notably, the twigs packed in vacuum-sealed aluminum-foil bags and cryopreserved using the CP3 method achieved the highest viability. In the current study, the sections were desiccated to a moisture content of 25–30% in a refrigerator at −5 °C. Pre-freezing was conducted using a CryoMed controlled rate freezer, with a cooling rate of 0.1 °C·min$^{-1}$ to −30 °C, followed by a 24-h hold at −30°C. The temperature was then decreased at a rate of 0.5 °C/min to −80 °C, and subsequently at 1 °C/min to −180 °C (CP3) in the same CryoMed controlled rate freezer (Forma 7451TF, Thermo Forma). After this process, the twigs were submerged in LN for 28 days. This procedure aligned with the three-step desiccation process utilized in *Malus* [58], which also indicates that desiccation to a 30% moisture content in a cold chamber at −5 °C is effective. The pre-freezing phase involved a controlled rate freezer with a cooling rate of 1 °C·h$^{-1}$ to −30 °C. The 24-h preservation at −30 °C provided efficient cold hardening. To enhance recovery, the treatment of artificial cold acclimation following collection is likely to benefit the buds' cold acclimation, thereby improving their resistance to cryopreservation. Sakai conducted classic research that identified a temperature range of −3 to −5 °C as the most effective for maximizing frost hardiness in twigs [59]. Vogiatzi et al. reported that the desiccation step at −4 °C was essential, and that the 24-h holding period at −30 °C contributed to improved recovery [13].

The distinct thawing methods can significantly influence the recovery of buds. In this study, slow thawing, which involved immersing the twigs of *V. amurensis* in water for 24 h at 4 °C in a refrigerator before transferring them to a substrate, resulted in markedly better recovery compared to those that were rapidly thawed. This difference may be attributed to the fact that slow thawing gradually restores the water absorption capacity of DBs, which lose excessive water during the desiccation process. Slow thawing has been widely utilized in *Malus* species. The cryopreserved tissues of *Malus* germplasm were subjected to gradual thawing through a nearly 24-h transfer to 4 °C in a cold room prior to removing the sections from vials and placing them in wet, autoclaved sand at the same temperature for 15 days of dehydration [38,58]. The researchers conducted a series of rehydration time tests on the preservation of different genotypes [55], and the results indicated that the genotypes exhibited varying requirements for thawing time. Therefore, the thawing duration for different genotypes of *V. amurensis* needs to be further adjusted and investigated.

Genotype and its response to the environment are equally important factors in cryopreservation. The genotype may influence the recovery of DBs following cryopreservation. Vogiatzi et al. demonstrated that the viability of winter-dormant apple buds after cryopreservation is strongly genotype-dependent [59]. Höfer conducted a detailed analysis of data from various *Malus* germplasms, indicating that the variability within a specific species deviated significantly from the standard deviations [60]. Genotypic differences in freezing tolerance were not reflected in the moisture content of the trees [61]. Reed showed that there was some genotypic variation in recovery; however, the viability of all genotypes exceeded the minimum threshold of 40% proposed as an acceptable limit for germplasm conservation [62]. Towill et al. adjusted the system to enhance the successful application of cryopreservation on dormant *Malus* buds, considering factors such as tolerance, moderation, and sensitivity [63]. In this study, the viabilities of the three species we harvested for cryopreservation all exceeded 55%. Therefore, the cryopreservation technique established in this research could be applied to the conservation of *V. amurensis* resources.

Dormancy, defined as the quiescent status of tissues and organs, may be relevant for cryopreservation due to its physiological basis related to growth systems and mechanisms of cold hardiness. Cold hardiness in woody perennials is influenced by a complex interplay of factors including the timing and rate of cold acclimation as well as the maximum level of cold tolerance achieved [64]. Furthermore, winter conditions prior to scion cutting significantly affect the moisture content and distribution within bud tissues, thereby impacting the recovery rate of DBs after cryopreservation [65]. Bilavcik et al. reported a direct correlation between cryosurvival and cold hardening [66]. The dormancy phase and extreme cold conditions exert notable effects on the post-cryopreservation survival of dormant *Malus* buds [13,39,56,59,67]. In this study, the viability of DBs of *V. amurensis* collected in November (autumn) and March (spring) was significantly lower compared to buds harvested in winter (January, February, and December). The majority of buds collected in November likely lacked sufficient dormancy to withstand cryopreservation. In contrast, buds collected in March had already exited their dormancy period, making them unable to survive the extremely low temperatures of LN. The duration of dormancy varies among different plant species across distinct regions. For dormant buds, cryopreservation is primarily suitable for species undergoing a dormancy phase, with low-temperature periods in the Northern Hemisphere typically occurring from late November to the end of January, occasionally extending into early February [68]. The timing for the successful cryopreservation of hybrid aspen aligns with findings that the maximum cold hardiness of forest trees in Finland is achieved between November and January [69]. Over 66% of DBs from *P. tremula*, originating from both coastal and continental Norway, have been shown to survive immersion in LN when collected in January and February [48]. In *P. tremuloides*, the cryopreservation of DBs collected from the mountains of Colorado, USA, demonstrated good survival rates of 79–92% in November and 100% in December [36]. Different from other grape varieties, *V. amurensis* is the most cold-resistant grape, so further research is needed to access the effectiveness of the proposed protocol on other grapevine species before its implementation.

The ultrastructure of cells in DBs of *V. amurensis* preserved at ultra-low temperatures was examined. It was observed that following the dehydration of meristematic cells, the degree of vacuolation decreased, along with a reduction in the content of free water within the cells. The ultrastructural changes in dehydrated cells serve as indicators of enhanced freeze resistance and dehydration tolerance, suggesting that these ultrastructural alterations are correlated with physiological changes, especially the water content and tissue [70]. Dehydration is a critical step in cryopreservation; only by removing sufficient water can cells achieve vitrification without the formation of ice crystals during rapid freezing. The cell membrane is a primary indicator of cellular damage during freezing, and electrolyte leakage serves as a measure of membrane integrity. Therefore, maintaining the integrity of the cell membrane is essential for maximizing the recovery rate of frozen samples. After cryopreservation, the perinuclear space expands, leading to significant nuclear deformation, severe contraction of the protoplasm, extensive damage to various organelles and membrane structures, and potentially lethal effects on the cells. Additionally, most of the cells experienced dehydration, which significantly enhanced their freeze resistance. Ultrastructural observations revealed that while the cell structure was slightly deformed, the organelles and membranes remained largely intact. Following thawing, the cells exhibited high viability, indicating that this cryopreservation method is effective for preserving the DBs of *V. amurensis*.

## 5. Conclusions

This is the first detailed report on the cryopreservation of DBs of *V. amurensis*, to the best of our knowledge. An improved recovery rate of 66.7% from the initial 30% was achieved through the program as follows. The bud viability was reached at a moisture content range of 25–30%. Then, the temperature of the twigs was decreased at a rate of 0.1 °C/min to −30 °C, maintaining this temperature for 24 h, followed by a decline of 0.5 °C/min to −80 °C, then a decline of 1 °C/min to −180 °C, before finally reaching −196 °C in a CryoMed controlled rate freezer. The low-temperature thawing procedure consisted of immersing the twigs in ice-cold (4 °C) water for 24 h and changing the water every hour during the initial 6 h to facilitate recovery. Twigs packed in aluminum-foil bags resulted in significantly higher recovery rates. Regarding the cell ultrastructure, the most significant changes in cells occurred during the processes of dehydration, freezing, and thawing. Therefore, optimizing the preservation procedure to achieve complete vitrification and enhance the relative viability of preserved cells is crucial for vitrification cryopreservation technology. In summary, it is only through a comprehensive investigation of the mechanisms underlying freezing injury and cellular antifreeze responses that we can establish a set of universally effective and straightforward cryopreservation methods for plant tissues and cells. This advancement will enable cryopreservation technology for plant cells to play a more significant role in bioengineering research and applications.

**Author Contributions:** Conceptualization, writing—original draft preparation, D.S. and P.N.; methodology, J.L.; formal analysis, Z.W.; investigation, G.S.; data curation, M.L.; writing—review and editing, X.Z.; supervision, project administration, funding acquisition, J.A. All authors have read and agreed to the published version of the manuscript.

**Funding:** This work was supported by the Talent Introduction Fund of Jilin Agricultural University (grant number 0214–202022920) and the Science and Technology Development Plan of Jilin Province (grant number 20240303112NC).

**Data Availability Statement:** The authors will supply the relevant data in response to reasonable requests.

**Conflicts of Interest:** The authors declare that the research was conducted in the absence of any commercial or financial relationships that could be construed as potential conflicts of interest.

## References

1. Engelmann, F.; Engels, J.M.M. Technologies and strategies for ex situ conservation. In Proceedings of the Managing Plant Genetic Diversity, Proceedings of an International Conference, Kuala Lumpur, Malaysia, 12–16 June 2000; pp. 89–103. [CrossRef]
2. Lambardi, M.; De Carlo, A.; Benelli, C.; Bartolini, G. Cryopreservation of woody species by vitrification of shoot tips and embryogenic tissue. *Acta Hortic.* **2001**, *560*, 125–128. [CrossRef]
3. Reed, B.M. Implementing cryopreservation for long-term germplasm preservation in vegetatively propagated species. In *Biotechnology in Agriculture and Forestry*; Towill, L.E., Bajaj, Y.P.S., Eds.; Cryopreservation of Plant Germplasm II; Springer: Berlin/Heidelberg, Germany, 2002; Volume 50, pp. 22–33. [CrossRef]
4. Hao, Y.-J.; Deng, X.-X. Genetically stable regeneration of apple plants from slow growth. *Plant Cell Tissue Organ Cult. (PCTOC)* **2003**, *72*, 253–260. [CrossRef]
5. Höfer, M. Conservation strategy of genetic resources in strawberry in Germany. *Acta Hortic.* **2011**, *918*, 139–146. [CrossRef]
6. Kovalchuk, I.; Lyudvikova, Y.; Volgina, M.; Reed, B.M. Medium, container and genotype all influence in vitro cold storage of apple germplasm. *Plant Cell Tissue Organ Cult. (PCTOC)* **2008**, *96*, 127–136. [CrossRef]
7. Lambardi, M.; Roncasaglia, R.; Previati, A.; De Carlo, A.; Dradi, G.; Da Re, F.; Calamai, L. In vitro slow growth storage of fruit rootstocks inside gastight or gas-permeable containers. *Acta Hortic.* **2006**, *725*, 483–488. [CrossRef]
8. Normah, M.N.; Chin, H.F.; Reed, B.M. *Conservation of Tropical Plant Species*; Springer Science & Business Media: New York, NY, USA, 2013; pp. 3–26. [CrossRef]
9. Reed, B.M. Plant cryopreservation: A continuing requirement for food and ecosystem security. *Vitr. Cell. Dev. Biol.—Plant* **2017**, *53*, 285–288. [CrossRef]
10. Reed, B.M. Cryopreservation—Practical considerations. In *Plant Cryopreservation: A Practical Guide*; Reed, B.M., Ed.; Springer: New York, NY, USA, 2008; pp. 3–13. [CrossRef]
11. Bettoni, J.C.; Bonnart, R.; Volk, G.M. Challenges in implementing plant shoot tip cryopreservation technologies. *Plant Cell Tissue Organ Cult. (PCTOC)* **2021**, *144*, 21–34. [CrossRef]
12. Vollmer, R.; Villagaray, R.; Egusquiza, V.; Espirilla, J.; García, M.; Torres, A.; Rojas, E.; Panta, A.; Barkley, N.A.; Ellis, D. The Potato Cryobank at The International Potato Center (Cip): A Model for Long Term Conservation of Clonal Plant Genetic Re-sources Collections of the Future. *Cryo Lett.* **2016**, *37*, 318–329. [PubMed]
13. Vogiatzi, C.; Grout, B.; Toldam-Andersen, T.; Green, J. Cryopreservation of dormant buds from temperate fruit crops to optimise working collection resources. *Acta Hortic.* **2011**, *908*, 477–482. [CrossRef]
14. Matsumoto, T.; Sakai, A. Cryopreservation of axillary shoot tips of in vitro-grown grape (Vitis) by a two-step vitrification protocol. *Euphytica* **2003**, *131*, 299–304. [CrossRef]
15. Matsumoto, T.; Sakai, A. Cryopreservation of in vitro-cultured axillary shoot tips of *Vitis* by vitrification. *Acta Hortic.* **2000**, *539*, 177–181. [CrossRef]
16. Sakai, A. Survival of the Twig of Woody Plants at −196 °C. *Nature* **1960**, *185*, 393–394. [CrossRef]
17. Sakai, A. Development of cryopreservation techniques. In *Cryopreservation of Tropical Plant Germplasm*; Engelmann, F., Takagi, H., Eds.; JIRCAS: Tsukuba, Japan, 2000; pp. 1–7.
18. Ryynänen, L. Viability and regeneration of dormant silver birch buds stored at super-low temperatures. *Can. J. For. Res.* **1996**, *26*, 617–623. [CrossRef]
19. Towill, L.E.; Forsline, P.L. Cryopreservation of sour cherry (*Prunus cerasus* L.) using a dormant vegetative bud method. *Cryo Lett.* **1999**, *20*, 215–222.
20. Jenderek, M.M.; Tanner, J.D.; Ambruzs, B.D.; West, M.; Postman, J.D.; Hummer, K.E. Twig pre-harvest temperature significantly influences effective cryopreservation of Vaccinium dormant buds. *Cryobiology* **2016**, *74*, 154–159. [CrossRef]

21. Seufferheld, M.J.; Stushnoff, C.; Forsline, P.L.; Gonzalez, G.H.T. Cryopreservation of cold-tender apple germplasm. *J. Am. Soc. Hortic. Sci.* **1999**, *124*, 612–618. [CrossRef]
22. Harvengt, L.; Meier-Dinkel, A.; Dumas, E.; Collin, E. Establishment of a cryopreserved gene bank of European elms. *Can. J. For. Res.* **2004**, *34*, 43–55. [CrossRef]
23. Volk, G.M.; Bonnart, R.; Waddell, J.; Widrlechner, M.P. Cryopreservation of dormant buds from diverse Fraxinus species. *Cryo Lett.* **2009**, *30*, 262–267.
24. Suzuki, M.; Niino, T.; Akihama, T.; Oka, S. Shoot formation and plant regeneration of vegetative pear buds cryopreserved at −150 degree. *J. Jpn. Soc. Hortic. Sci.* **1997**, *66*, 29–34. [CrossRef]
25. Kuoksa, T.; Hohtola, A. Freeze-preservation of buds from scots pine trees. *Plant Cell Tissue Organ Cult. (PCTOC)* **1991**, *27*, 89–93. [CrossRef]
26. Wang, Z.X.; Jiao, Z.Q.; Xu, P.L.; Chen, L.; Ai, J.; Liu, X.M.; Yang, Y.M. Bisexual flower ontogeny after chemical induction and berry characteristics evaluation in male *Vitis amurensis* Rupr. *Sci. Hortic.* **2013**, *162*, 11–19. [CrossRef]
27. Bak, M.-J.; Truong, V.L.; Kang, H.-S.; Jun, M.; Jeong, W.-S. Anti-inflammatory effect of procyanidins from wild grape (*Vitis amurensis*) seeds in LPS-induced RAW 264.7 cells. *Oxidative Med. Cell. Longev.* **2013**, *2013*, 409321. [CrossRef]
28. Peng, S.-C.; Cheng, C.-Y.; Sheu, F.; Su, C.-H. The antimicrobial activity of heyneanol A extracted from the root of Taiwanese wild grape. *J. Appl. Microbiol.* **2008**, *105*, 485–491. [CrossRef] [PubMed]
29. Lee, H.-R.; Bak, M.-J.; Jeong, W.-S.; Kim, Y.-C.; Chung, S.-K. Antioxidant properties of proanthocyanidin fraction isolated from wild grape (*Vitis amurensis*) seed. *J. Korean Soc. Appl. Biol. Chem.* **2009**, *52*, 539–544. [CrossRef]
30. Min, Z.; Guo, Z.; Wang, K.; Zhang, A.; Li, H.; Fang, Y. Antioxidant effects of grape vine cane extracts from different Chinese grape varieties on edible oils. *Molecules* **2014**, *19*, 15213–15223. [CrossRef] [PubMed]
31. Jeong, H.Y.; Kim, J.Y.; Lee, H.K.; Ha, D.T.; Song, K.-S.; Bae, K.; Seong, Y.H. Leaf and stem of *Vitis amurensis* and its active components protect against amyloid β protein (25–35)-induced neurotoxicity. *Arch. Pharmacal Res.* **2010**, *33*, 1655–1664. [CrossRef] [PubMed]
32. Lee, E.O.; Lee, H.J.; Hwang, H.S.; Ahn, K.S.; Chae, C.; Kang, K.S.; LU, J.; Kim, S. Potent inhibition of Lewis lung cancer growth by heyneanol A from the roots of *Vitis amurensis* through apoptotic and anti-angiogenic activities. *Carcinogen* **2006**, *27*, 2059–2069. [CrossRef]
33. Song, R.G.; Zheng, Y.C.; Lu, W.P.; Shen, Y.J.; Fan, S.T.; Yang, Y.M.; Lin, X.G. Genetic analysis of resistance to Plasmopara viticola disease in progenies (F_1-F_4) derived from interspecific hybridization with *Vitis amurensis*. *J. Fruit Sci.* **2008**, *25*, 33–39.
34. Ge, Y.X.; Shen, Y.J.; Li, X.H.; Ai, J.; Zang, P. Vistas in the research of the major economic characters of the germplasm resources of *Vitis amurensis*. *Spec. Wild Econ. Animal Plant Res.* **1997**, *4*, 37–42.
35. Bettoni, J.C.; Marković, Z.; Bi, W.; Volk, G.M.; Matsumoto, T.; Wang, Q.-C. Grapevine Shoot Tip Cryopreservation and Cryotherapy: Secure Storage of Disease-Free Plants. *Plants* **2021**, *10*, 2190. [CrossRef] [PubMed]
36. Bettoni, J.C.; Kretzschmar, A.A.; Bonnart, R.; Shepherd, A.; Volk, G.M. Cryopreservation of 12 Vitis Species Using Apical Shoot Tips Derived from Plants Grown In Vitro. *HortScience* **2019**, *54*, 976–981. [CrossRef]
37. Bi, W.-L.; Hao, X.-Y.; Cui, Z.-H.; Volk, G.M.; Wang, Q.-C. Droplet-vitrification cryopreservation of in vitro-grown shoot tips of grapevine (Vitis spp.). *Vitr. Cell. Dev. Biol.—Plant* **2018**, *54*, 590–599. [CrossRef]
38. Esensee, V.; Stushnoff, C.; Forsline, P.L. Cryopreservation of dormant grape (*Vitis* sp.) buds. *HortScience* **1990**, *25*, 1090e. [CrossRef]
39. Forsline, P.L.; Towill, L.E.; Waddell, J.W.; Stushnoff, C.; Lamboy, W.F.; McFerson, J.R. Recovery and longevity of cryopreserved dormant apple buds. *J. Am. Soc. Hortic Sci.* **1998**, *123*, 365–370. [CrossRef]
40. Towill, L.E.; Ellis, D.D. Cryopreservation of dormant buds. In *Plant Cryopreservation: A Practical Guide*; Reed, B.M., Ed.; Springer: New York, NY, USA, 2008; pp. 421–422.
41. Parfitt, D.E.; Almehdi, A.A. Cryogenic storage of grape pollen. *Am. J. Enol. Vitic.* **1983**, *34*, 227–228. [CrossRef]
42. Ganeshan, S. Cryogenic preservation of grape (*Vitis vinifera* L.) pollen. *Vitis* **1985**, *173*, 169–173.
43. Hassan, N.A.; Gomma, A.H.; Shahin, M.A.; El Homosany, A.A. In vitro storage and cryopreservation of some grape varieties. *J. Hortc. Sci. Ornam. Plants* **2013**, *5*, 183–193. [CrossRef]
44. Wang, Q.; Mawassi, M.; Sahar, N.; Li, P.; Violeta, C.-T.; Gafny, R.; Sela, I.; Tanne, E.; Perl, A. Cryopreservation of grapevine (*Vitis* spp.) embryogenic cell suspensions by encapsulation-vitrification. *Plant Cell Tissue Organ Cult. (PCTOC)* **2004**, *77*, 267–275. [CrossRef]
45. Wang, Q.; Gafny, R.; Sahar, N.; Sela, I.; Mawassi, M.; Tanne, E.; Perl, A. Cryo-preservation of grapevine (*Vitis vinifera* L.) embryogenic cell suspensions and subse-quent plant regeneration by encapsulation-dehydration. *Plant Sci.* **2002**, *162*, 551–558. [CrossRef]
46. Dussert, S.; Mauro, M.C.; Deloire, A.; Hamon, A.; Engelmann, F. Cryopreserva-tion of grape embryogenic cell suspensions 1: Influence of pretreatment, freezing and thawing conditions. *Cryo Lett.* **1991**, *12*, 287–298.
47. Dussert, S.; Mauro, M.C.; Engelmann, F. Cryopreservation of grape embryogenic cell suspensions 2: Influence of post-culture conditions and application to different strains. *Cryo Lett.* **1992**, *13*, 15–22.
48. Carra, A.; Carimi, F.; Bettoni, J.C.; Pathirana, R. Progress and Challenges in the Application of Synthetic Seed Technology for Ex Situ Germplasm Conservation in Grapevine (*Vitis* spp.). In *Synthetic Seeds*; Faisal, A., Alatar, A.A., Eds.; Springer Nature: Cham, Switzerland, 2019; pp. 439–467. [CrossRef]

49. Stushnoff, C.; Junttila, O. Seasonal development of cold stress resistance in several plant species at a coastal and a continental location in North Norway. *Polar Biol.* **1986**, *5*, 129–133. [CrossRef]
50. Jokipii, S.; Ryynänen, L.; Kallio, P.T.; Aronen, T.; Häggman, H. A cryopreservation method maintaining the genetic fidelity of a model forest tree, *Populus tremula* L. × *Populus tremuloides* Michx. *Plant Sci.* **2004**, *166*, 799–806. [CrossRef]
51. Lambardi, M.; Fabbri, A.; Caccavale, A. Cryopreservation of white poplar (*Populus alba* L.) by vitrification of in vitro-grown shoot tips. *Plant Cell Rep.* **2000**, *19*, 213–218. [CrossRef] [PubMed]
52. Tsvetkov, I.; Benelli, C.; Capuana, M.; De Carlo, A.; Lambardi, M. Application of vitrification-derived cryotechniques for long-term storage of poplar and aspen (*Populus* spp.) germplasm. *Agric. Food Sci.* **2009**, *18*, 160–166. [CrossRef]
53. Volk, G.M.; Waddell, J.; Bonnart, R.; Towill, L.; Ellis, D.; Luffman, M. High viability of dormant *Malus* buds after 10 years of storage in liquid nitrogen vapour. *Cryo Lett.* **2008**, *29*, 89–94. [CrossRef]
54. Vogiatzi, C.; Grout, B.W.W.; Wetten, A.; Toldam-Andersen, T.B. Cryopreservation of winter-dormant apple buds: I. Variation in recovery with cultivar and winter conditions. *Cryo Lett.* **2011**, *32*, 358–366. [CrossRef]
55. Höfer, M. Cryopreservation of winter-dormant apple buds: Establishment of a duplicate collection of Malus germplasm. *Plant Cell Tissue Organ Cult. (PCTOC)* **2015**, *121*, 647–656. [CrossRef]
56. Towill, L.E.; Bonnart, R. Cryopreservation of apple using non-desiccated sections from winter-collected scions. *Cryo Lett.* **2005**, *26*, 323–332.
57. Rantala, S.; Kaseva, J.; Karhu, S.; Veteläinen, M.; Uosukainen, M.; Häggman, H. Cryopreservation of *Ribes nigrum* (L.) dormant buds: Recovery via in vitro culture to the field. *Plant Cell Tissue Organ Cult. (PCTOC)* **2019**, *138*, 109–119. [CrossRef]
58. Sakai, A. Studies of frost hardiness in woody plants II. Effect of temperature on hardening. *Plant Physiol.* **1966**, *41*, 353–359. [CrossRef] [PubMed]
59. Vogiatzi, C.; Grout, B.; Wetten, A. Cryopreservation of winter-dormant apple: Iii—Bud water status and viability after cooling to −30 °C and during recovery from cryopreservation. *Cryo Lett.* **2012**, *33*, 160–168.
60. Höfer, M.; Hanke, M.-V. Cryopreservation of fruit germplasm. *Vitr. Cell. Dev. Biol.—Plant* **2017**, *53*, 372–381. [CrossRef]
61. Aronen, T.; Ryynanen, L. Cryopreservation of dormant in vivo-buds of hybrid aspen: Timing as a critical factor. *Cryo Lett.* **2014**, *35*, 385–394.
62. Reed, B.M.; Denoma, J.; Luo, J.; Chang, Y.; Towill, L. Cryopreservation and long-term storage of pear germplasm. *Vitr. Cell. Dev. Biol.—Plant* **1998**, *34*, 256–260. [CrossRef]
63. Towill, L.E.; Forsline, P.L.; Walters, C.; Waddell, J.W.; Laufmann, M.J. Cryopreservation of *Malus* germplasm using a winter vegetative bud method: Results from 1915 accessions. *Cryo Lett.* **2004**, *25*, 323–334. [CrossRef]
64. Rowland, I.J.; Ogden, E.L.; Ehlenfeldt, M.K.; Arora, R. Cold tolerance of blueberry genotypes throughout the dormant period from acclimation to deacclimation. *HortScience* **2008**, *43*, 1970–1974. [CrossRef]
65. Vogiatzi, C.; Grout, B.W.W.; Wetten, A.; Toldam-Andersen, T. Critical steps in cryopreservation of dormant winter buds collected under relatively mild winter conditions. *Cryobiology* **2010**, *61*, 368. [CrossRef]
66. Bilavcik, A.; Zamecnik, J.; Faltus, M. Cryotolerance of apple tree bud is independent of endodormancy. *Front. Plant Sci.* **2015**, *6*, 695. [CrossRef] [PubMed]
67. Tyler, N.; Stushnoff, C. Dehydration of dormant apple buds at different stages of cold acclimation to induce cryopreservibility in different cultures. *Can. J. Plant Sci.* **1988**, *68*, 1169–1176. [CrossRef]
68. Jenderek, M.M.; Reed, B.M. Cryopreserved storage of clonal germplasm in the USDA national plant germplasm system. *Vitr. Cell. Dev. Biol.—Plant* **2017**, *53*, 299–308. [CrossRef]
69. Koski, V. The timing of hardening and dehardening of forest trees. *Acta Hortic.* **1985**, *168*, 117–124. [CrossRef]
70. Ishikawa, M.; Ide, H.; Price, W.S.; Arata, Y.; Nakamura, T.; Kishimoto, T. Freezing behaviours in plant tissues: Visualization using NMR micro-imaging and biochemical regulatory factors involved. In *Plant Cold Hardiness: From the Laboratory to the Field*; CABI: Wallingford, UK, 2009; pp. 19–28. [CrossRef]

*plants*

MDPI

*Article*

# Integrating Physiology, Transcriptome, and Metabolomics Reveals the Potential Mechanism of Nitric Oxide Concentration-Dependent Regulation of Embryo Germination in *Sorbus pohuashanensis*

**Caihong Zhao [1], Yue Zhang [1] and Ling Yang [1,2,*]**

[1] State Key Laboratory of Tree Genetics and Breeding, Northeast Forestry University, Harbin 150040, China; zhaocaihong@nefu.edu.cn (C.Z.); zhangy-dl@nefu.edu.cn (Y.Z.)
[2] College of Forestry, Beijing Forestry University, Beijing 100091, China
* Correspondence: yangling2024@bjfu.edu.cn

**Abstract**

Nitric oxide (NO) breaks a seed's dormancy and stimulates germination by signaling. However, the key physiological metabolic pathways and molecular regulatory mechanisms are still unclear. Therefore, this study used physiological, transcriptomic, and metabolomics methods to analyze the key genes and metabolites involved in the NO regulation of plant embryo germination and their potential regulatory mechanisms. The physiological analysis results indicate that the appropriate concentration of NO increased the content of NO and hydrogen peroxide ($H_2O_2$) in cells, stimulated the synthesis of ethylene and jasmonic acid (JA), induced a decrease in abscisic acid (ABA) content, antagonistic to the gibberellin ($GA_3$) effect, and promoted embryo germination and subsequent seedling growth. However, the high concentrations of NO caused excessive accumulation of $H_2O_2$, destroyed the reactive oxygen species (ROS) balance, and inhibited embryo germination and seedling growth. The combined analysis of transcriptomics and metabolomics showed that the genes related to phenylpropanoid (phenylalanine ammonia-lyase, trans-cinnamate 4-monooxygenase, ferulate-5-hydroxylase, coniferyl-alcohol glucosyltransferase), and flavonoid synthesis (10 genes such as *CHS*) were significantly up-regulated during embryo germination. The high concentration of exogenous NO inhibited embryo germination by up-regulating the expression of 4-coumaric acid coenzyme A ligase (4CL) and negatively regulating the expression of flavonoid synthesis genes. This suggests that NO concentration-dependently regulates phenylpropanoid and flavonoid biosynthesis, thereby affecting ROS metabolism and hormone levels, and ultimately regulates the dormancy and germination of *Sorbus pohuashanensis* embryos.

**Keywords:** nitric oxide; seed dormancy; *Sorbus pohuashanensis*; transcriptome; metabolome; hormones

## 1. Introduction

*Sorbus pohuashanensis* has a high ornamental value [1]. Its seeds have a deep dormancy [2]. Sodium nitroprusside (SNP) can release NO, which is a potent plant regulatory substance that affects the content and metabolism of ROS and NO in cells under light conditions [3–5]. A large number of studies have reported that exogenous NO donor-SNP can relieve seed dormancy and promote germination by increasing NO levels [6–8].

However, higher concentrations of exogenous NO may affect the biosynthesis pathway of phenylpropanoids by inhibiting the expression and enzyme activity of some key genes, thereby inhibiting seed germination [9]. Therefore, the concentration of NO is one of the key factors affecting seed dormancy and germination.

NO regulates seed germination with the production of reactive oxygen species (ROS) and reactive nitrogen species (RNS), which is a necessary signal from dormancy to germination [10]. Many studies have confirmed that NO may have dual functions. For example, low concentrations of NO break the embryo dormancy of *S. pohuashanensis (Sorbus pohuashanensis)* by relying on ethylene synthesis [6]. This hormone-related signaling pathway is closely related to the accumulation of reactive oxygen species and antioxidant defense responses [11,12]. However, a high concentration of NO has a complex effect on seed germination, which can not only promote germination and enhance stress resistance but also may inhibit seed germination due to the high concentration. In addition, the promotion or inhibition of seed germination by NO is related to the role of ROS metabolism-related enzymes [13]. For instance, peroxisomes can produce NO and participate in lipid metabolism, amino acid oxidation, and ROS metabolism [14]. Catalase (CAT) can be activated by phosphorylation, which enhances its catalytic ability to scavenge ROS [15]. The physiological response of seed germination, which is regulated by NO, can be explained by the changes in the transcriptional abundance of enzymatic and non-enzymatic antioxidant defense genes [16,17]. Furthermore, the response and adaptation of plant embryos to NO is a very complex process involving the regulation of numerous genes in different metabolic pathways. NO and ROS can act independently in the same signaling pathway and have similar downstream responses and outcomes, but their internal molecular regulatory mechanisms need to be further investigated [18].

In this work, we used RNA sequencing (RNA-seq) and UPLC-MS/MS technology to provide an integrated map of transcriptomics and metabonomics of *S. pohuashanensis* embryos to fill the gaps in understanding the gene expression and metabolic regulation during the dormancy release and germination of *S. pohuashanensis* embryos caused by exogenous NO. We have provided a large body of evidence that the regulation of NO on *S. pohuashanensis* embryo germination is activated by a combination of gene transcription and metabolic homeostasis, particularly through the regulation of reactive nitrogen species (RNS), ROS, and hormones. These results help us to understand the mechanism of NO concentration-dependent regulation of embryo germination in *S. pohuashanensis*. It provides a reference for the study of dormancy release and plant seed germination at the multi-omics level.

## 2. Results

### 2.1. Phenotypic Concentration-Dependent Response to NO on Embryo Germination of *S. pohuashanensis*

To help understand how SNP affects the growth of *S. pohuashanensis* embryos, we used a growth model in which *S. pohuashanensis* embryos were cultured in 5 mL of a sodium nitroprusside solution (0~100 mmol/L) for 8 days. We used a growth model to culture the embryos of *S. pohuashanensis* in 5 mL SNP if solution (0~100 mmol/L) for 8 days. The results showed that SNP had a biphasic concentration effect on the dormancy release of *S. pohuashanensis* seeds. Among them, compared with the control, the 40 mmol/L and 60 mmol/L SNP treatments significantly increased the embryo germination rate of *S. pohuashanensis* (increased by 28.9% and 39.2%, respectively), while 80 mmol/L and 100 mmol/L SNP treatments decreased by 6.2% and 9.3%, respectively (Figure 1).

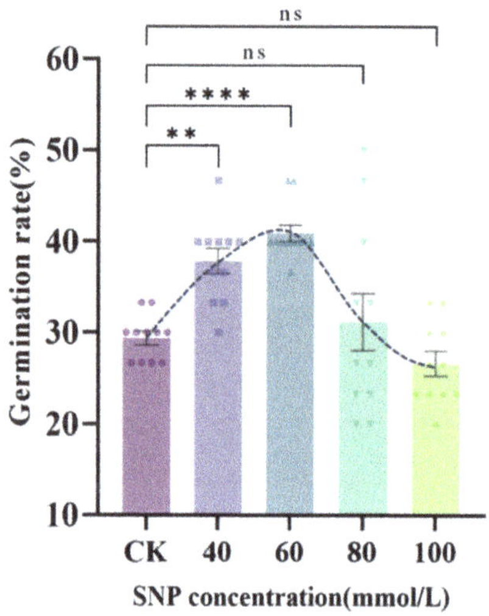


**Figure 1.** Effect of different concentrations of SNP on the embryo germination rate of *Sorbus pohuashanensis*. (Note: ns means $p > 0.05$, ** means $p \leq 0.01$, **** means $p \leq 0.0001$).

At the same time, we also found that, compared with untreated (CK, water), the embryos of *S. pohuashanensis* treated with 60 mmol/L of SNP (PG for short) germinated faster, the seedlings grew healthily, and the leaves developed well. The embryos of *S. pohuashanensis* treated with 100 mmol/L of SNP (IG for short) germinated faster in the pre-germination stage, and the seedlings grew poorly and yellowed in the mid-germination stage (Figure 2a). In this study, the germination process was divided into three stages (Figure 2b): pre-germination (0–3 d), mid-germination (3–5 d), and late-germination (5–8 d). During the pre-germination stage, the two cotyledons began to stretch slightly and turned pale green. At the mid-germination stage, the color of the two cotyledons continues to turn green, the opening angle of the cotyledons increases, accompanied by the elongation of the hypocotyl, and the radicle begins to appear. At the late-germination stage, the hypocotyls and radicles elongated rapidly.

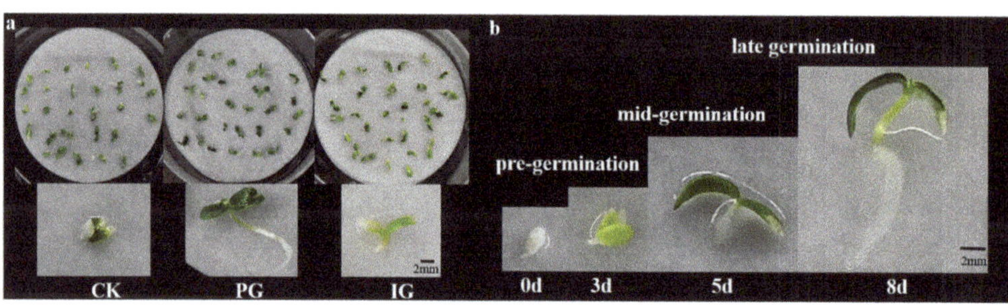


**Figure 2.** Germination model of an embryo of *Sorbus pohuashanensis*. (**a**) The phenotype of *Sorbus pohuashanensis* embryos cultured to day 8 after treatment with different SNP concentrations; (**b**) There are three stages of embryo germination of *Sorbus pohuashanensis*: 0–3 d is pre-germination, 3–5 d is mid-germination, and 5–8 d is late-germination.

## 2.2. Physiological Response to NO in Embryo Germination of S. pohuashanensis

### 2.2.1. The Response of NO-Dependent Concentration to RNS During the Embryonic Germination of S. pohuashanensis

To investigate the dynamic changes in physiological responses of *S. pohuashanensis* embryos during germination, NO content, nitric oxide synthase (NOS), and nitrate reductase activities (NR) were measured at 3 h, 3 d, 5 d, and 8 d after germination. The endogenous NO content increased continuously during the germination of the control embryos (Figure 3a, CK). At the pre-germination and mid-germination stages, the endogenous NO content of the embryos increased after treatment with exogenous NO (PG and IG) (compared to CK) but decreased at the late-germination stage (Figure 3a). At the pre-germination stage (3 h–3 d), PG treatment significantly increased endogenous NOS (Figure 3b) and NR (Figure 3c) activities (compared to CK), and thereafter NOS and NR decreased (but not less than the control). After 3 h of IG treatment, NOS activity was significantly higher than that of CK, and thereafter, there was no significant difference from CK (Figure 3b), while NR activity always showed no significant difference from CK (Figure 3c).

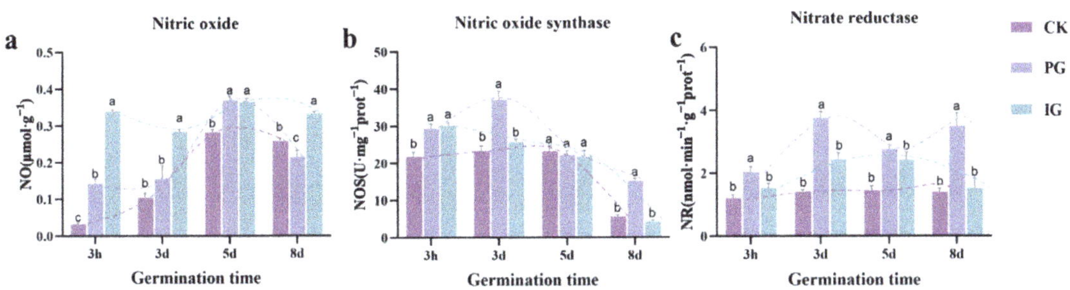


**Figure 3.** Effect of different concentrations of SNP treatment on the RNA content of *Sorbus pohuashanensis* embryos. (**a**) NO content. (**b**) NOS. (**c**) NR. Note: Different small letters in the figure indicate significant differences between the different treatments ($p < 0.05$).

### 2.2.2. The Response of NO-Dependent Concentration to ROS During the Embryonic Germination of S. pohuashanensis

To investigate the dynamic changes of ROS during germination of the embryo of *S. pohuashanensis* under different concentrations of NO, the levels of $H_2O_2$ and superoxide anions, as well as the activities of the corresponding enzymes, were measured during germination. The content of $H_2O_2$ and superoxide anion in CK and PG first increased and then decreased, with PG and IG significantly increasing the content of $H_2O_2$ and decreasing the content of superoxide anion in the pre-germination stage (Figure 4a) and the content of $H_2O_2$ and superoxide anion in IG was significantly higher than that in CK in the late-germination stage.

During pre-germination, PG significantly increased the activities of enzymes related to reactive oxygen species, such as SOD, POD, and CAT. During the late-germination stage, the activities of POD and CAT were significantly higher in PG-treated embryos than those of CK and IG. In addition to affecting the activity of active oxygen-scavenging enzymes, NO treatment also increased the levels of antioxidant substances (GSH and GSSG) and showed a " first increasing and then decreasing" trend in the germination process.

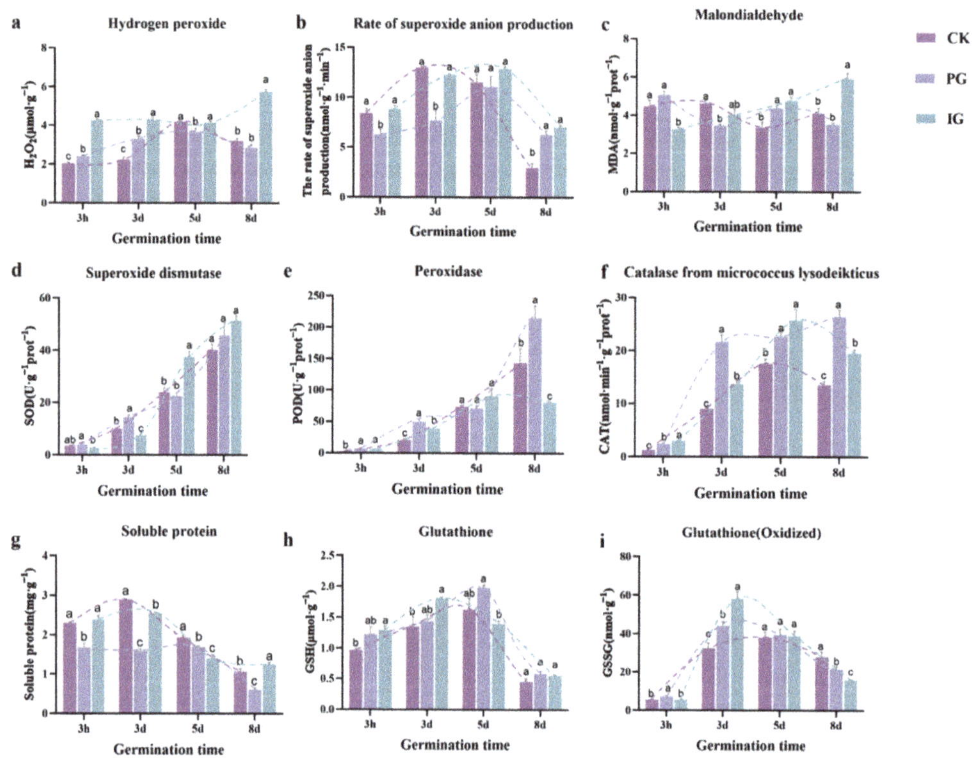


**Figure 4.** Effect of different SNP concentrations on the content of reactive oxygen species in the embryos of *Sorbus pohuashanensis*. (**a**) $H_2O_2$ content. (**b**) superoxide anion formation rate. (**c**) malondialdehyde content. (**d**) SOD content. (**e**) POD content. (**f**) CAT content. (**g**) soluble protein content. (**h**) glutathione (**i**) glutathione (oxidized). Note: Different small letters in the figure indicate significant differences between the different treatments ($p < 0.05$).

### 2.2.3. The Concentration-Dependent Hormone Response to NO on Embryo Germination of *S. pohuashanensis*

The levels of ABA, $GA_3$, ACC, and JA were determined to investigate the concentration-dependent regulation of the hormone balance during germination of the embryo *S. pohuashanensis* (Figure 5). The ABA content of embryos treated with PG and IG was higher than that of the control. $GA_3$/ABA and JA/ABA of Ig treatment were significantly higher than those of CK and PG. With increasing germination time, JA/ABA of IG treatment first decreased and then increased, while CK and PG first increased and then decreased (consistent with the biphasic concentration–response curve).

The results of the correlation analysis showed (Figure 6) that the NO content of *S. pohuashanensis* embryos had a significant positive correlation with $H_2O_2$ and JA and a significant negative correlation with ABA. There was a significant positive correlation between NO content and $H_2O_2$ content in *S. pohuashanensis* embryos. $H_2O_2$ was positively correlated with JA, $GA_3$, and ACC and negatively correlated with ABA.

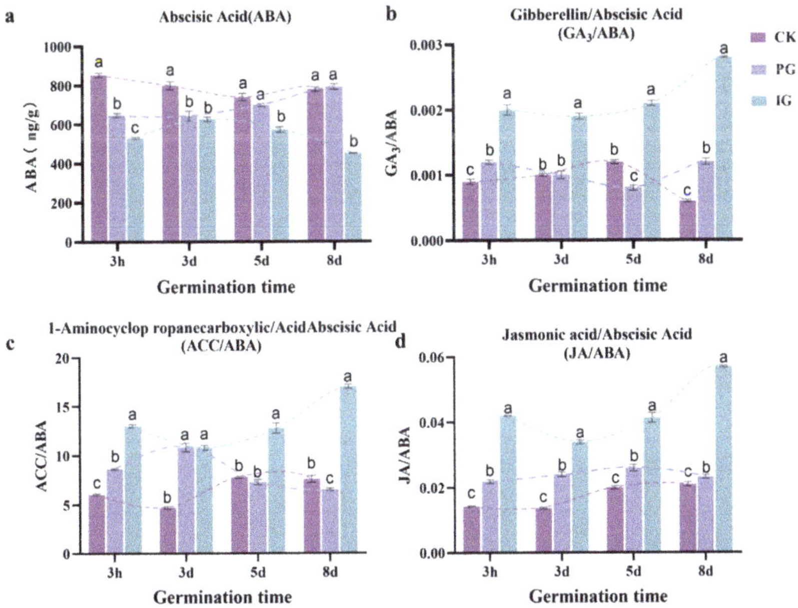


**Figure 5.** Effect of different concentrations of SNP treatment on the hormone content of *Sorbus pohuashanensis* embryos. (**a**) ABA content. (**b**) $GA_3$/ABA. (**c**) ACC/ABA. (**d**) JA/ABA. Note: Different small letters in the figure indicate significant differences between the different treatments ($p < 0.05$).

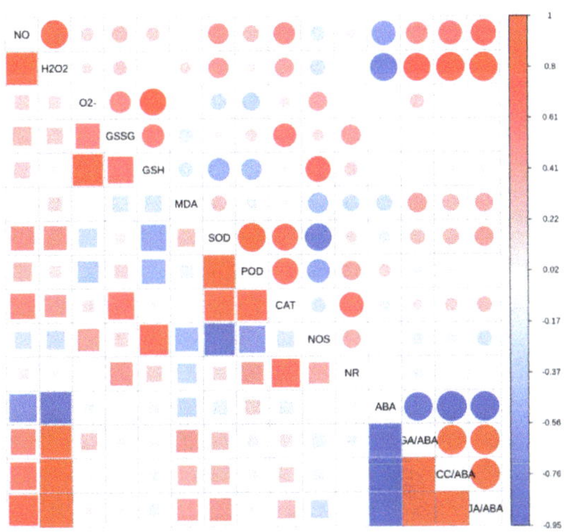


**Figure 6.** Correlation analysis of RNS, ROS, and hormones. Note, the diagonal represents the sample name. The color from blue to red indicates the change of correlation from negative to positive. The larger the figure of the upper and lower triangular regions, the greater the correlation, and vice versa.

### 2.3. Summary of Transcriptome Analysis

In this study, we selected embryos treated with CK, PG, and IG at 3 h and 3 d in the pre-germination stage as experimental materials for transcriptome sequencing analysis. A total of 43,304 differentially expressed genes were identified by RNA-seq analysis (clean reads; q30 $\geq$ 91.76%; see Table 1 for a summary of RNA sequencing data). The GC contents of these sequences were higher than 45.96%, indicating that the sequencing results were of high quality. By pairwise comparison, the differentially expressed genes DEGs were determined between exogenous NO treatments with different concentrations (Figure 7a). Compared to CK-3h, PG-3h and IG-3h up-regulated 2247 and 1355 genes and down-regulated 3083 and 2440 genes, respectively. Compared to CK-3d, PG-3d and IG-3d up-regulated 85 and 183 genes and down-regulated 911 and 143 genes, respectively. Interestingly, the number of differentially expressed genes was significantly higher 3 h after embryo germination was significantly higher than after 3 days. All DEGs are divided into two categories, one grouped by the DEGs in 3 h and the other by the DEGs in 3 d (Figure 7b).

**Table 1.** The RNA-seq information of samples.

| Sample | Raw Reads | Clean Reads | Q20 (%) | Q30 (%) | GC Content (%) |
|---|---|---|---|---|---|
| CK-3h-1 | 56,725,510 | 55,160,662 | 97.82 | 93.14 | 46.34 |
| CK-3h-2 | 54,750,690 | 53,237,138 | 97.89 | 93.37 | 46.41 |
| CK-3h-3 | 53,956,558 | 52,231,748 | 97.9 | 93.39 | 46.29 |
| CK-3d-1 | 64,708,806 | 62,885,714 | 97.9 | 93.37 | 46.8 |
| CK-3d-2 | 70,509,068 | 68,250,268 | 97.84 | 93.18 | 46.83 |
| CK-3d-3 | 62,635,840 | 60,728,020 | 98 | 93.72 | 46.79 |
| PG-3h-1 | 58,613,942 | 56,862,708 | 97.8 | 93.06 | 46.1 |
| PG-3h-2 | 80,612,590 | 78,953,442 | 98.48 | 95.86 | 45.96 |
| PG-3h-3 | 62,406,360 | 60,066,066 | 97.92 | 93.44 | 46.08 |
| PG-3d-1 | 51,615,176 | 50,409,014 | 97.35 | 91.76 | 46.67 |
| PG-3d-2 | 55,948,152 | 54,195,682 | 97.99 | 93.67 | 46.77 |
| PG-3d-3 | 58,913,916 | 57,146,268 | 97.87 | 93.3 | 46.8 |
| IG-3h-1 | 63,467,176 | 61,660,990 | 97.93 | 93.47 | 46.07 |
| IG-3h-2 | 65,084,926 | 63,275,686 | 97.79 | 93.06 | 46.15 |
| IG-3h-3 | 75,233,136 | 72,775,194 | 97.82 | 93.09 | 46.23 |
| IG-3d-1 | 67,369,076 | 65,026,244 | 97.83 | 93.15 | 46.79 |
| IG-3d-2 | 58,658,806 | 56,981,024 | 97.79 | 93.05 | 46.73 |
| IG-3d-3 | 61,788,888 | 60,412,016 | 98.01 | 93.71 | 46.81 |

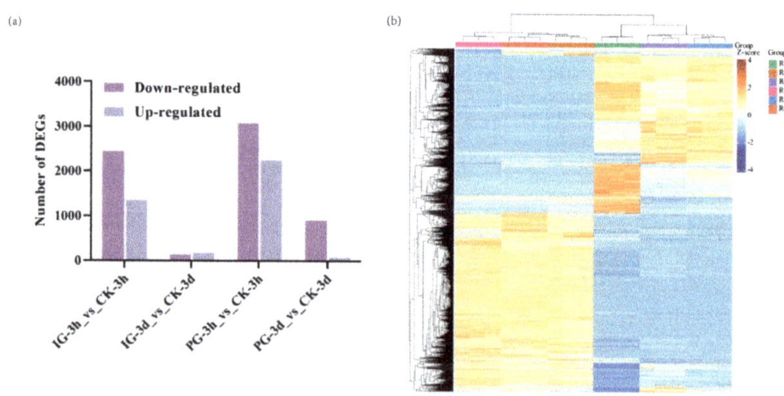


**Figure 7.** Effects of different concentrations of NO treatment on the transcriptome changes of *Sorbus pohuashanensis* embryos. (**a**) The number of differential genes in *Sorbus pohuashanensis*. (**b**) Differential enrichment analysis of *Sorbus pohuashanensis*.

## *2.4. Differentially Expressed Genes Analysis*

The results of KEGG and GO enrichment showed that the non-overlapping DEGs of NO concentration-dependently promoted the embryonic germination of *S. pohuashanensis* and were significantly enriched in lipid synthesis and metabolism, carbohydrate metabolism, amino acid biosynthesis, and metabolism, while the non-overlapping DEGs of NO-concentration-dependently inhibited the embryonic germination of *S. pohuashanensis* and were significantly enriched in plant hormone signal transduction, flavonoid biosynthesis, amino acid metabolism, and other pathways (Supplementary Material Figures S1 and S2). In PG-3d compared to CK-3d and IG-3d compared to CK-3d, the NO-dependent, non-overlapping DEGs that promoted embryo germination were significantly enriched in the MAPK signaling pathway, protein processing in the endoplasmic reticulum and plant hormone signaling, while non-overlapping DEGs that inhibited embryo germination were significantly enriched in fructose and mannose metabolism, pentose and glucuronate interconversions, plant hormone signaling transduction, and other pathways.

In summary, we found that exogenous NO can promote seed germination by regulating lipid metabolism, sugar conversion, and carbohydrate metabolism. The DEGs that inhibit germination regulate embryo germination mainly via plant hormone signaling, sugar conversion, and other metabolic pathways.

### 2.5. Validation of Differentially Expressed Genes by qRT-PCR

There was no significant difference between RNA-Seq and qRT-PCR data of *S. pohuashanensis*, and the trend was similar. Therefore, it was confirmed that the RNA-Seq data used in this study were reliable (Figure 8).

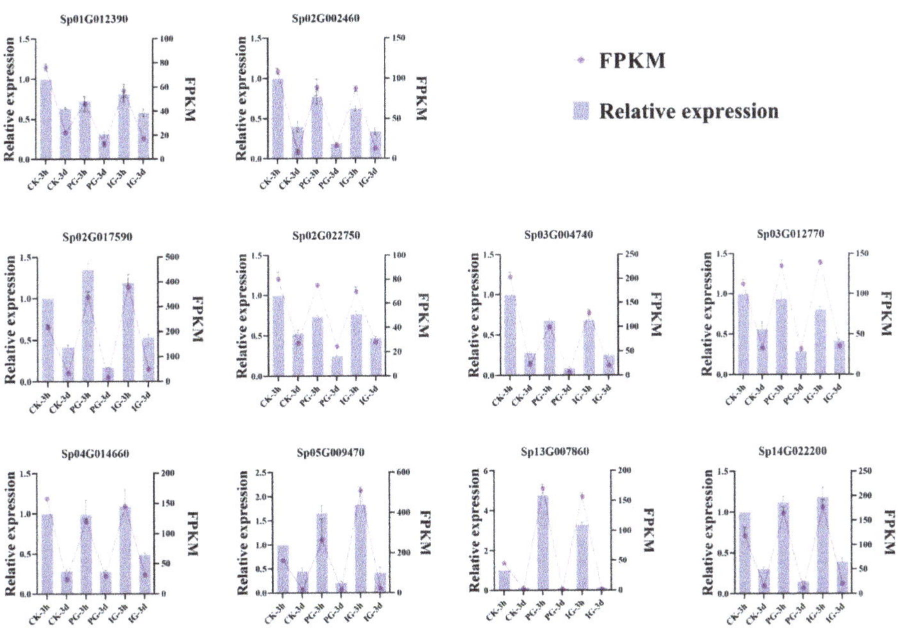


**Figure 8.** Verification of differential genes by qRT-PCR.

### 2.6. Metabolome Analysis

Using metabolomics analysis, the PCA method was used to compare the metabolite peaks detected between 3 h and 3 d after different concentrations of SNP treatment (Figure 9), and samples were separated into PC1 (58.16%) and PC2 (6.81%). The levels of metabolites in the samples of 3 h and 3 d were separated, indicating that there was a significant difference in metabolism between the samples of 3 h and 3 d. The metabolites of PG, IG, and CK in the 3 d samples were separated from each other, but the content of metabolites in the 3 h samples was summarized, indicating that the germination of seeds at the 3 d level changed significantly at the metabolic level. This shows that the transient effect of NO makes the change in transcriptome faster than the change in metabolite level.

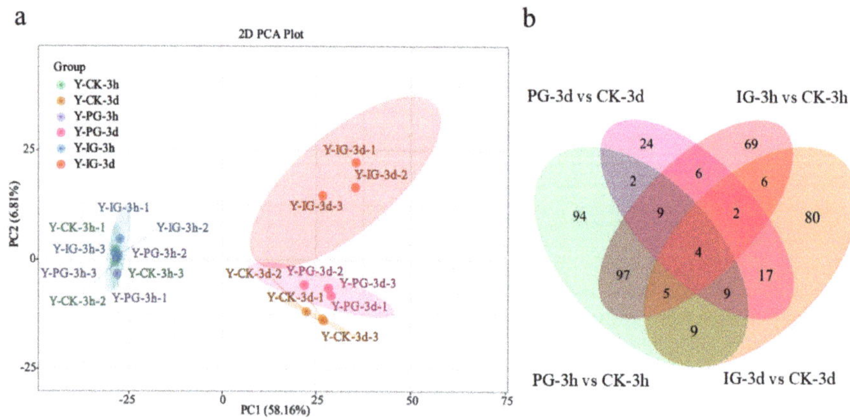


**Figure 9.** (**a**) PCA scores were detected in *Sorbus pohuashanensis* embryos at 3 h and 3 d after different concentrations of SNP treatment. PCA scores were derived from all metabolites detected in the three replicate samples under each treatment. (**b**) Venn diagram of DAMs among various pair-wise comparisons.

Subsequently, the differentially expressed metabolites were identified by comparing three replicate samples treated with different SNP concentrations for 3 h and 3 d (Figure 10). A total of 229 metabolites (120 up-regulated and 109 down-regulated) were screened between PG-3h and CK-3h, 73 metabolites (34 up-regulated and 39 down-regulated) were screened between PG-3d and CK-3d, 198 metabolites (96 up-regulated and 102 down-regulated) were screened between IG-3h and CK-3h, and 132 metabolites (98 up-regulated and 34 down-regulated) between IG-3d and CK-3d.

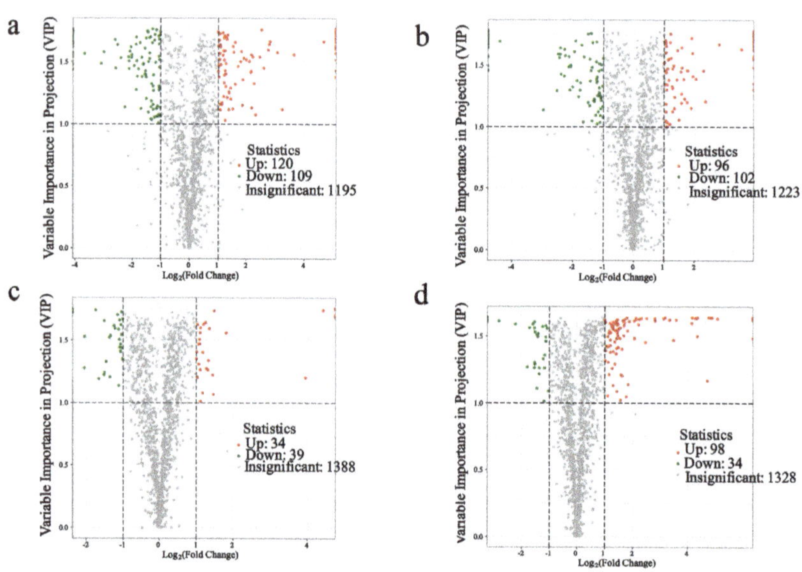


**Figure 10.** Difference analysis of gene expression after treatment with different concentrations of SNP. (**a**) PG-3h vs. CK-3h; (**b**) PG-3d vs. CK-3d; (**c**) IG-3h vs. CK-3h; (**d**) IG-3d vs. CK-3d.

To analyze the function of DAMs, the major differentially expressed metabolites of NO that concentration-dependently regulate the dormancy of *S. pohuashanensis* embryos for 3 h, we compared the differentially expressed metabolites of PG vs. CK and IG vs. CK (Supplementary Material Figure S3). In the non-overlapping-differential metabolites of PG-3h vs. CK-3h and IG-3h vs. CK-3h, the differential metabolites of NO concentration-dependently promoted embryo germination of *S. pohuashanensis* and were mainly enriched in lipid metabolism, biosynthesis of secondary metabolites and biosynthesis of amino acids. The differential metabolites of NO that concentration-dependently inhibited the embryo germination of *S. pohuashanensis* were mainly enriched in the biosynthesis of flavone and flavonol biosynthesis, amino acid metabolism, and carbohydrate pathways. It is hypothesized that low concentrations of exogenous NO may promote the germination of *S. pohuashanensis* embryos through the biosynthesis of amino acids. If the concentration of exogenous NO is too high, it may inhibit embryo germination through amino acid metabolism. This is consistent with the results of transcriptomics. In the non-overlapping differential metabolites of PG-3d vs. CK-3d and IG-3d vs. CK-3d, the differential metabolites of NO that concentration-dependently promoted the embryo germination of *S. pohuashanensis* were mainly enriched in the biosynthesis of secondary metabolites, while the differential metabolites of NO the concentration-dependently inhibited the embryo germination of *S. pohuashanensis* were mainly enriched in lipid metabolism, biosynthesis of secondary metabolites, amino acid metabolism, and other metabolic pathways. By analyzing different metabolites between 3 h and 3 d, the results showed that for embryos treated with NO for 3 h, DAMs were significantly enriched in phenylpropane and flavonoid biosynthesis, lipid metabolism, and phenylalanine metabolism (also in a concentration-dependent manner). The DAMs treated for 3 days were significantly enriched in flavonoid and flavonol biosynthesis, phenylalanine metabolism, and other metabolic pathways.

These results suggest that changes in metabolite accumulation are strictly controlled by differential gene expression. Flavonoid biosynthesis and phenylpropanoid biosynthesis are considered to be the major metabolic pathways of exogenous NO affecting embryo germination. Therefore, we analyzed the metabolites and crucial genes involved in NO-related metabolic pathways and investigated the mechanism of NO's concentration-dependent effect on dormancy selection of dormancy and germination of *S. pohuashanensis* embryos.

### *2.7. Comprehensive Metabolome and Transcriptome Analysis of Differential Regulation of Key Genes and Metabolites*

To further analyze the DEGs and DAMs, we annotated them in the KEGG database and found many joint common pathways (Figure 11). PG-3h vs. CK-3h was enriched in 45 pathways, while IG-3h vs. CK-3h was enriched in 9 pathways. Among these pathways, the flavonoid biosynthetic pathway and the phenylpropanoid biosynthetic pathway were highly enriched pathways under PG and IG treatments. The correlation coefficients of the nine quadrants showed that the differential expression of genes and metabolites was consistent in the third and seventh quadrants (Figure 12). We conclude that these genes are positively correlated with the regulation of metabolites and that the changes in metabolites may be positively regulated by these genes.

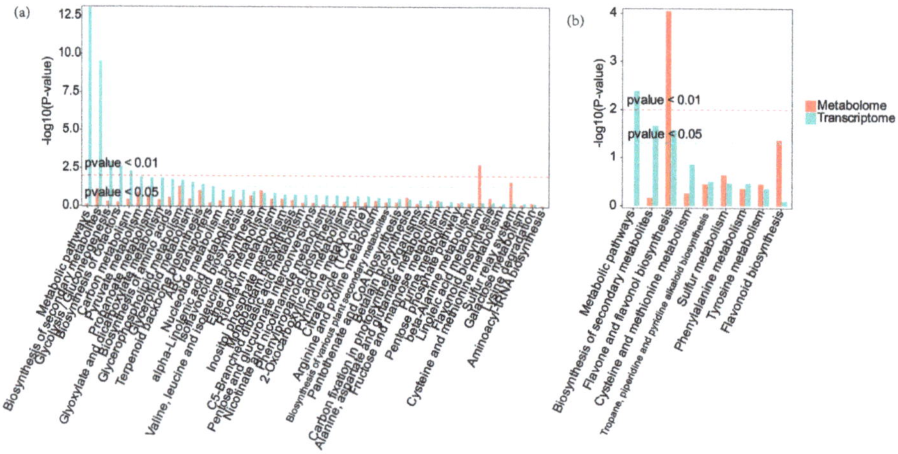


**Figure 11.** KEGG pathways of DEGS and DAMs. (**a**) PG-3h vs. CK-3h. (**b**) IG-3h vs. CK-3h.

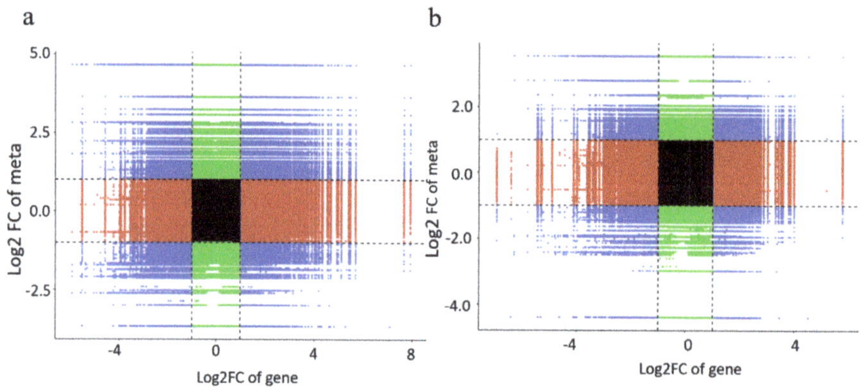


**Figure 12.** Transcriptome and metabolome combined analysis of nine quadrants. The black dotted line, from left to right, from top to bottom, is divided into 1-9 quadrants. (**a**) PG-3h vs. CK-3h, (**b**) IG-3h vs. CK-3h.

### *2.8. The Role of Differential Genes and Differential Metabolites in Phenylpropanoid Biosynthesis and Flavonoid Biosynthesis Pathways*

Considering all the results of the transcriptome and metabolome, we focused on the genes and metabolites that showed significant differences between PG-3h and IG-3h. The results showed that phenylpropanoid and flavonoid genes/metabolites changed significantly after treatment with different concentrations of exogenous NO. We hypothesize that these genes and metabolites may play a different role in the response of PG and IG to pre-embryonic germination.

Therefore, we performed a Pearson correlation analysis on 450 genes and 30 metabolites in the phenylpropanoid and flavonoid biosynthetic pathways and constructed a correlation network diagram of the different genes and the different metabolites. Based on $|r| > 0.8$, $p < 0.05$ as the threshold, the regulatory co-expression network was visualized in Cytoscape (Figure 13). The results show that a total of 390 nodes were connected to 1365 edges of PG-3h, and 405 nodes were connected to 1109 edges of IG-3h. In the comparison between PG-3h and CK-3h, there were 62 key gene nodes: 5 key metabolite nodes significantly associated with phenylpropanoid biosynthesis, 11 key gene nodes, 6 key metabolite nodes associated with flavonoid biosynthesis, and 11 key gene nodes involved in both phenylpropanoid biosynthesis and flavonoid biosynthesis. In IG-3h compared to CK-3h, there were 47 key gene nodes: 4 key metabolite nodes related to the phenylpropanoid biosynthetic pathway, 21 key gene nodes, 6 key metabolite nodes involved in flavonoid biosynthesis pathway, and 6 key gene nodes related to both the phenylpropanoid and flavonoid biosynthetic pathway. We hypothesize that these genes and metabolites may play a key role in the response of exogenous NO's concentration-dependent effects on the release of dormancy in *S. pohuashanensis* embryos.

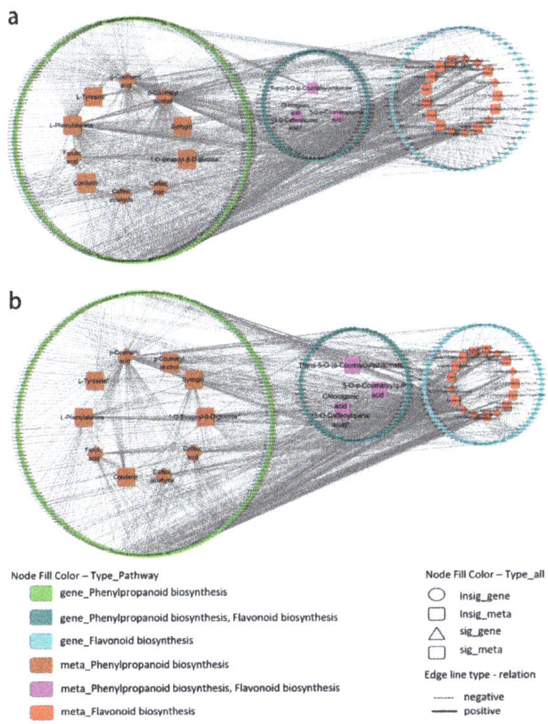


**Figure 13.** Regulatory co-expression network of different genes and different metabolites related to the biosynthesis of phenylpropanoids and flavonoids. The Cytoscape software (version 3.10.0) was used to visualize the network. (**a**) PG-3h vs. CK-3h, (**b**) IG-3h vs. CK-3h.

Based on the expression patterns of key genes and metabolites, we constructed models for the phenylpropanoid and flavonoid biosynthetic pathways. About 93 key DEGs and 5 DAMs were involved in the phenylpropanoid biosynthetic pathway, and about 37 key DEGs and 6 DAMs were involved in the flavonoid biosynthetic pathway. About 68 major DEGs and 4 DAMs were involved in the phenylpropanoid biosynthetic pathway, and about 27 major DEGs and 6 DAMs were involved in the flavonoid biosynthetic pathway.

Under exogenous NO treatment, a large number of phenylpropanoid biosynthesis genes were upregulated, and their expression levels were significantly higher than those of the control (Figure 14). Compared with CK treatment, PG significantly promoted phenylalanine ammonia-lyase [EC: 4.3.1.24] and trans-cinnamate 4-monooxygenase [EC: 1.14.14.91] in the phenylpropanoid biosynthetic pathway. The expression of ferulate5-hydroxylase [EC: 1.14.-.-] and coniferyl-alcohol glucosyltransferase [EC: 2.4.1.111] was significantly inhibited by IG treatment, while the expression of 4-coumarate-CoA ligase [EC: 6.2.1.12] and coniferyl-alcohol glucosyltransferase [EC: 2.4.1.111] was significantly inhibited by IG treatment. The key metabolites, such as p-coumaric acid, caffeic acid, ferulic acid, and caffeoyl-aldehyde, were down-regulated overall. Therefore, different concentrations of exogenous NO treatment may continue to affect the metabolism of phenylpropanoids. The higher the concentration of exogenous NO, the lower the expression of the metabolites.

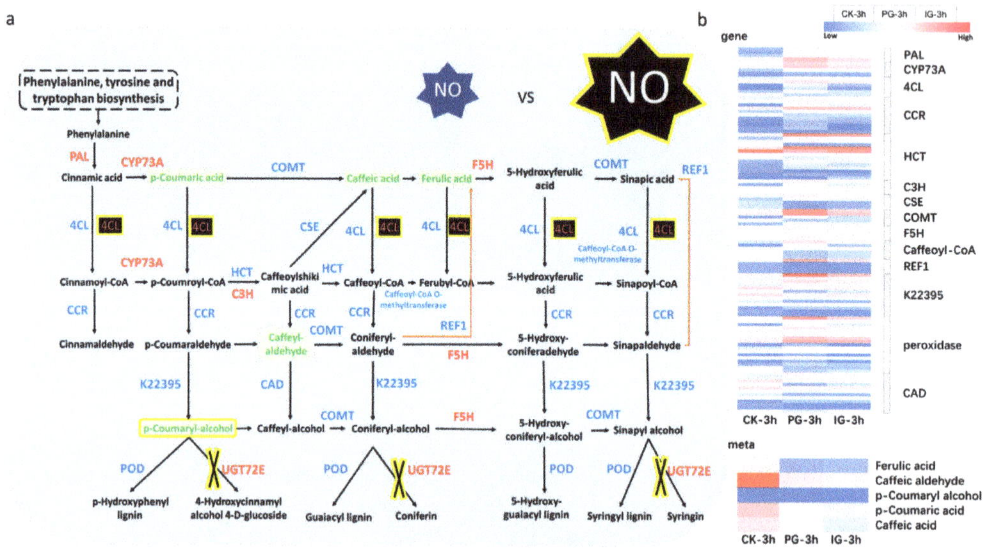


**Figure 14.** Phenylpropanoid biosynthetic pathway. (**a**) NO concentration-dependent regulation of phenylpropanoid biosynthesis signaling pathway during embryo germination of *Sorbus pohuashanensis*. The blue frame indicates PG-3h vs. CK-3h, and the yellow frame indicates that IG-3h vs. CK-3h influences the pathway of phenylpropanoid biosynthesis compared to PG-3h vs. CK-3h. Red indicates upregulated genes/metabolites, blue indicates upregulated and downregulated genes/metabolites, and green indicates downregulated genes/metabolites. (**b**) Heatmap of gene and metabolite expression levels. Note, PAL, phenylalanine ammonia-lyase; CYP73A, trans-cinnamate 4-monooxygenase; 4CL, 4-coumarate-CoA ligase; CCR, cinnamoyl-CoA reductase; HCT, shikimate O-hydroxycinnamoyltransferase; CSE, caffeoylshikimate esterase; COMT, caffeic acid 3-O-methyltransferase/acetylserotonin O-methyltransferase; CYP98A(C3′H), 5-O-(4-coumaroyl)-D-quinate 3′-monooxygenase; K22395, cinnamyl-alcohol dehydrogenase; POD, peroxidase; UGT72E, coniferyl-alcohol glucosyltransferase; CAD, cinnamyl-alcohol dehydrogenase; REF1, coniferyl-aldehyde dehydrogenase; CYP84A (F5H), ferulat-5-hydroxylase.

Compared to CK-treated embryos, most of the genes involved in flavonoid biosynthesis were upregulated by exogenous NO (Figure 15). With the increase in NO concentration, the expression levels of key genes such as trans-cinnamate 4-monooxygenase, 5-O-(4-coumaroyl)-D-quinate 3′-monooxygenase, flavonol synthase, and anthocyanidin reductase were decreased. This shows that exogenous NO upregulates the up-regulation of flavonoid synthesis genes, but the higher the concentration, the lower the series of flavonoid synthesis genes.

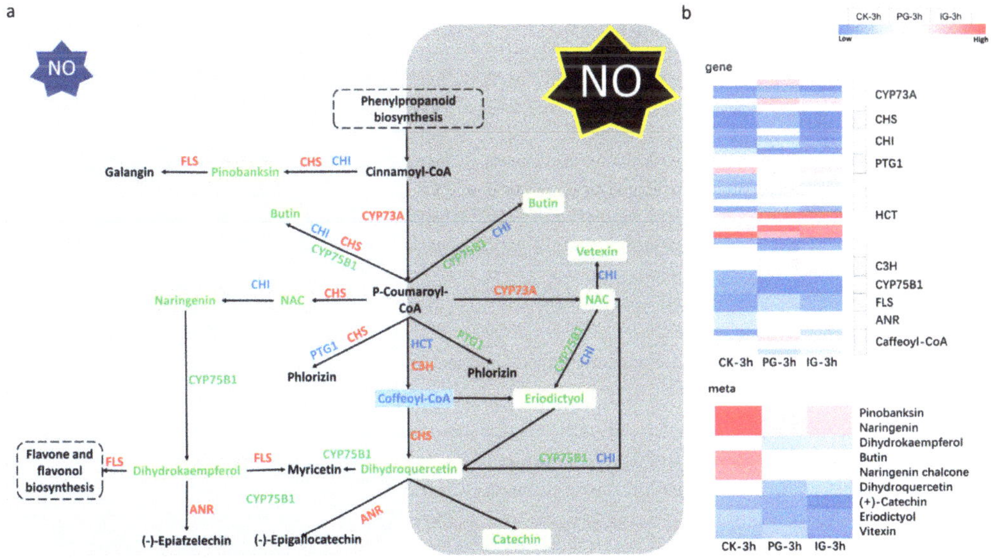


**Figure 15.** The pathway of flavonoid biosynthesis pathway. (**a**) NO concentration-dependent regulation of flavonoid biosynthesis signaling pathway during embryo germination of *Sorbus pohuashanensis*. The blue outer frame indicates PG-3h compared to CK-3h, and the grey outer frame indicates IG-3h compared to CK-3h. (**b**) Heatmap of gene and metabolite expression levels. Note, CHI, chalcone isomerase; CHS, chalcone synthase, FLS, flavonol synthase; NAC, naringenin chalcone; CYP75B1, flavonoid 3′-monooxygenase; CYP73A, trans-cinnamate 4-monooxygenase; HCT, shikimate O-hydroxycinnamoyl transferase; CYP98A(C3′H), 5-O-(4-coumaroyl)-D-quinate 3′-monooxygenase; ANR, anthocyanidin reductase; PTG1, phlorizin synthase.

## 3. Discussion

### 3.1. NO Concentration-Dependently Regulates ROS Signaling During Embryo Dormancy Release in S. pohuashanensis

NO is regulated by both environmental and endogenous signaling and has both synergistic and antagonistic effects. It is a free radical that can effectively regulate physiological responses in plant growth and development [19–22]. Our results showed that PG could promote the embryo germination of *S. pohuashanensis*, but embryo germination and seedling growth and development were inhibited after IG treatment. This biphasic concentration effect is consistent with previous reports on seed germination in *Arabidopsis thaliana* (L.) Heynh. [4,23], apple (*Malus pumila* Mill.) [24], barley (*Hordeum vulgare* L.) [4], *Paulownia elongata* S. Y. Hu [25] and many halophytes [26]. At the same time, the NO content in *S. pohuashanensis* embryos increased after SNP treatment for 3 h, which may be due to the increase in endogenous NO level in embryos by exogenous NO through oxidation or reduction pathway [27] and the increase in NO content may come from enzyme or non-enzyme pathway [28]. The results of Krasuska showed that the transition of apple flowers from dormancy to germination is related to the increase in NOS activity [10]. This is in line with our results. Due to the high activity and instability of NO, the existence of NOS in the context of its synthesis in higher plants is very elusive. Moreover, it is not yet clear at the molecular level how NO concentration regulates embryonic dormancy release and germination [10]. In the stage of pre-germination, in this study, the decrease in nitrate reductase activity caused by high concentrations of nitric oxide may be due to the participation of NR in the synthesis of NO. Exogenous NO controls its production by regulating GSNOR activity and nitrate assimilation to reduce NR activity [19,29,30].

ROS (such as $H_2O_2$, superoxide anions, etc.) and NO play a crucial role in regulating the dormancy and germination of plant seeds [31]. Low levels of ROS act as signaling molecules that promote the abrogation of dormancy and trigger seed germination [8]. Excessive accumulation of ROS accumulation can lead to the accumulation of reactive oxygen species in germinated seeds, and the change in the redox state of cells leads to the induction of gene expression [5]. In our results, PG treatment led to a significant increase in hydrogen peroxide content in embryos during embryo dormancy release, which positively regulated embryo germination and seedling morphogenesis. With increasing culture duration, treatment with a high concentration of exogenous NO led to a significantly higher $H_2O_2$ content than in the control and PG treatment. In the late stage of germination (8 d), the excessive accumulation of $H_2O_2$ in IG-treated *S. pohuashanensis* embryos destroyed the balance of ROS and affected the growth and development of seedlings in the late stage of seed germination. The ROS content depends not only on the ROS production system but also on the efficiency of the redox feedback mechanism between NO and ROS content [31], such as SOD. Many studies have shown that the activity of antioxidant enzymes increases during seed impregnation [32]. However, in most cases, the activation of the antioxidant system is a late event in the germination process, and the activity of antioxidant enzymes increases in the presence of stress [15]. In our results, the significant increase in SOD, POD, and CAT levels caused by treatment with a high concentration of exogenous NO on day 5 of embryo germination confirmed this point, and the significant changes in POD and CAT activities at the early and late stages of germination depended on the concentration of exogenous NO [14]. It is characterized by the active metabolism of ROS and is a target for NO-mediated PTM [33]. Treatment with PG significantly increased the CAT activity of *S. pohuashanensis* embryos during germination. Compared with PG, the significant inhibition of CAT by IG might mean that the ability of embryos to remove internal $H_2O_2$ is lower, which is closely related to the physiological or adverse processes leading to an increase in oxidative metabolism [14,34]. The effect was significant during embryo dormancy (on the third day of germination). The effect was significant during the dormancy release period of the embryo (the third day of germination). This result is consistent with our previous research results, findings suggesting that NO acts upstream of $H_2O_2$, thereby regulating CAT activity [35]. Similar studies in plants have also shown that NO donors can react with superoxide anions and inhibit enzyme activity [14,36].

Studies have shown that the process of cell division depends on the increase in glutathione levels during the transition from germination to seedling growth [37–39]. This is consistent with our findings that the appropriate concentration of SNP treatment resulted in an increase in GSH levels during the abrogation of embryonic dormancy. However, the excessive accumulation of GSSG is toxic, and the high concentration of exogenous NO treatment leads to the inhibition of embryonic germination and seedling growth of *S. pohuashanensis*.

### *3.2. NO Concentration-Dependently Regulates Hormone Levels During Embryo Dormancy and Germination in S. pohuashanensis*

Plant hormones act together with the gas signaling molecules NO and ROS as second messengers to regulate seed germination and seedling growth after germination [5]. Seed germination and dormancy, although mainly dependent on hormone balance, are primarily regulated by abscisic acid, gibberellin, ethylene, and JA by maintaining appropriate ROS/RNS levels [40]. $GA_3$ and ABA are considered to be the most important hormones counteracting the regulation of seed germination [41]. Similar to previous studies, our results also showed that NO concentration regulated ABA content in the embryo throughout seed germination and seedling growth and development. NO significantly promoted embryo germination and seedling growth by causing a decrease in ABA content, which counteracted the effect of $GA_3$. Furthermore, the ABA content of the IG treatment was significantly lower than that of the control, while the $GA_3$ content was significantly higher than that of the control. Our previous study found that NO significantly abolished embryo dormancy in *S. pohuashanensis*, which was related to the increase in ethylene synthesis [35]. In this experiment, the concentration-dependent effect of NO on endogenous ethylene content was studied. The results showed that IG could significantly promote the increase in ethylene content in the embryo of *S. pohuashanensis*, which was significantly higher than that of PG treatment. At the same time, in the period of embryonic dormancy, release, seedling growth, and development were significantly higher than that of the control. In this study, it was found that treatment with exogenous NO throughout germination and seedling morphogenesis exogenous NO treatment significantly increased the endogenous JA content of *S. pohuashanensis* embryos, and the JA content increased with the increase in exogenous NO concentrations. This indicates that NO concentration-dependently stimulated the biosynthesis of JA, which was an important step in the abrogation of embryonic dormancy and late seedling growth and development of *S. pohuashanensis* by NO. It is well known that JA can regulate multiple adaptive responses of plants to various environmental stresses [42]. In this study, the increase in JA content caused by the high concentration of exogenous NO may be due to the stress response of plants to excessive exogenous NO. However, the JA-mediated protective mechanism of the embryo caused by exogenous NO needs to be further investigated. In addition, in agriculture, plant hormones can regulate the genes encoding NR, NIR, GS, and GOGAT and the activities of corresponding enzymes in the process of NO regulating the absorption and transport of nitrogen by plants, thereby regulating nitrogen assimilation [43].

### *3.3. NO Concentration-Dependently Regulated the Changes of Key Genes and Metabolites in the Biosynthesis Pathway of Phenylpropanoid and Flavonoid in the Early Stage of Embryo Germination of S. pohuashanensis*

We performed a comprehensive analysis of transcriptomes and metabolomes to determine the metabolites and metabolic pathways significantly affected by exogenous NO concentration-dependent regulation at the early stage of *S. pohuashanensis* embryo germination, especially phenylpropane and flavonoid biosynthesis related to seed germination [44]. Phenylpropanoids make an important contribution to NO regulation of *S. pohuashanensis* embryo germination. The flavonoids produced by phenylpropanoids have a strong antioxidant capacity and can scavenge ROS [45], protect seeds from oxidative damage, and play an important role in seed germination. Similarly, in our results, the differentially expressed genes in NO concentration-dependently promoted/inhibited embryo germination of *S. pohuashanensis* and were significantly enriched in the phenylpropanoid biosynthetic pathway. By comparing the nonoverlap of differential genes and differential metabolites in PG-3h vs. CK-3h and IG-3h vs. CK-3h, the specific metabolic pathways of NO's concentration-dependent regulation of *S. pohuashanensis* embryos were determined. Our results showed that the differential genes of NO concentration-dependently promoted Sorbus embryo germination and were significantly enriched in the phenylpropanoid and flavonoid biosynthetic pathways. The differential genes of NO that concentration-dependently inhibited embryo germination of *S. pohuashanensis* were significantly enriched in phenylalanine metabolism and flavonoid and flavonol biosynthesis. The phenylpropanoid biosynthetic pathway begins with phenylalanine metabolism, so a high concentration of exogenous NO in embryo treatment can inhibit the upstream pathway of the phenylpropanoid biosynthetic pathway, thereby inhibiting embryo germination.

The genes related to phenylpropanoid biosynthesis (PAL, CYP73A, F5H, UGT72E) were significantly upregulated by an appropriate concentration of exogenous NO (PG) during the regulation of embryo germination of *S. pohuashanensis*. PAL is the first step of the phenylpropanoid pathway involved in the synthesis of phenolic compounds, such as phenolic acids, lignin, flavonoids, etc. It plays an important role in the formation of the cell wall and the connection between cells in the process of exogenous NO regulation of germination of *S. pohuashanensis* embryos [46]. UGT72E is mainly involved in the glycosylation process of phenylpropanoid compounds. In this study, UGT72E was significantly upregulated under PG treatment, but there was no significant difference under IG treatment, suggesting that an appropriate concentration of exogenous NO may affect the structure and function of the cell wall by regulating the glycosylation process and promote the embryo germination of *S. pohuashanensis* embryo. 4CL (4-coumaric acid-CoA ligase) is an important gene in phenylpropanoid biosynthesis, which affects the biosynthesis of flavonoids and plays an important role in plant physiology and plant protection [46]. In this study, we found that the expression of 4CL was significantly upregulated during embryonic germination of *S. pohuashanensis*, which is regulated by a high concentration of NO, in contrast to an appropriate concentration of exogenous NO, and that a high concentration of exogenous NO can inhibit embryonic germination by impairing the high expression of 4CL. In conjunction with our previous studies, we found that exogenous NO could significantly increase the ROS content in *S. pohuashanensis* embryos, suggesting that the reduction of p-coumaric acid, caffeic acid, ferulic acid, and caffeyl-aldehyde can stimulate the accumulation of ROS in *S. pohuashanensis* embryos, thereby inhibiting the germination of *S. pohuashanensis* embryos treated with high concentrations of exogenous NO.

Flavonoids are widely distributed secondary metabolites. In this study, embryo germination was influenced by the concentration of key genes and metabolites involved in the flavonoid biosynthetic pathway. CHS is the first reaction step responsible for catalyzing the flavonoid pathway. It is significantly upregulated in NO, dependent on concentration, to promote *S. pohuashanensis* embryo germination, but the difference is not significant under IG treatment, suggesting that the appropriate concentration of exogenous NO promotes the accumulation of the CHS gene family to promote seed germination. Similar to our results, the abrogation of seed dormancy of *Polygonatum sibiricum* significantly promoted the germination of the flavonoid biosynthetic pathway [47]. Most of the flavonoid biosynthetic genes were upregulated in the embryos of *S. pohuashanensis* treated with exogenous NO (Figure 13). However, with increasing NO concentration, the expression of key genes such as CYP73A, C3H, FLS, and ANR decreased. It is suggested that although exogenous NO causes upregulation of flavonoid synthesis genes, the higher the concentration, the lower the expression of a number of flavonoid synthesis genes decreases as the concentration increases. Flavonoids and phenylpropanoids play an important role in the elimination of ROS [48,49]. In this study, the appropriate concentration of exogenous NO treatment could promote germination by upregulating the accumulation of flavonoids and regulating the ROS content, while the higher concentration of exogenous NO negatively regulates the synthesis of flavonoid synthesis genes, thereby inhibiting embryo germination. Interestingly, our analysis of the metabolic results showed that metabolites associated with this pathway were significantly downregulated.

## 4. Materials and Methods

### 4.1. Plant Material

In October 2018, ripe berries were collected from adult mother trees of *S. pohuashanensis* in Maoershan Experimental Forest Farm (127°30′–127°34′ E, 45°21′–45°25′ N) in Shangzhi City, Heilongjiang Province, and the seeds were processed by water selection [50]. The plump, pure, and ripe seeds with a harmless moisture content (9–10%) were put into plastic bags and stored at 0–5 °C. The weight of 1000 seeds, water content, and viability were measured before the experiment and then imbibed in distilled water at 20 °C ± 5 °C for

48 h. After imbibition, the seeds were stirred and soaked in 0.2% ($v/v$) NaCl solution for 10 min, washed with clean water, and the seed coat was peeled off on ice (*S. pohuashanensis* is an endosperm-free seed) to obtain the naked embryo out for the experiment [6].

*4.2. SNP Pretreatment and Embryo Germination Conditions*

A Petri dish with a diameter of 35 mm was placed in the large Petri dish, and a filter paper completely soaked with 5 mL of 0.05 mol·L$^{-1}$ Hepes-KOH (pH 7.0) was placed on top; 30 stripped embryos were placed on it. Then, the Petri dish with 5 mL of SNP solution (0, 40, 60, 80, or 100 mmol/L) was added to the same Petri dish, and, finally, the large Petri dish was sealed with a preservation film, and each treatment was repeated three times. Gaseous NO is released directly from the donor dish and acts on the embryos in the recipient dish. The embryos were irradiated with 25 °C and 60 μ mol·m$^{-2}$·S$^{-1}$ light for 3 h and germinated at16 h of light/8 h of darkness. The germination rate was calculated to determine the optimal SNP concentration after 8 days of germination in the culture room for 8 days. The standard for germination standard is that the cotyledons turn green, and the hypocotyl radicle elongation reaches 2 mm [5]. The calculation method of germination rate was based on the percentage of seed germination to the total number of tested seeds [6].

*4.3. Determination of Related Indexes of Active Nitrogen and Active Oxygen*

Tissue samples were taken 3 h and 1–8 days after SNP treatment. The relevant indicators of RNS are based on the method of Wang (2023) [6]. A total of 0.1 g fresh samples of embryos or seedlings were extracted and analyzed for RNS and ROS content.

*4.4. Determination of Hormone Content (ABA, GA$_3$, ACC, and JA)*

Hormone levels (ABA, GA$_3$, ACC, and JA) were detected by enzyme-linked immunosorbent assay (ELISA) with double antibody and one-step sandwich method. The absorbance (OD value) was determined at a wavelength of 450 nm, and the sample concentration was calculated [6,51].

*4.5. Transcriptome Analysis*

Total RNA was extracted from *S. pohuashanensis* embryos by ethanol precipitation and the CTAB-PBIOZOL method. After the library construction and detection were qualified, the DNB (DNA Nano Ball) was prepared and then loaded onto the sequencing chip to be sequenced with the high-throughput sequencer. The offline data was filtered to obtain clean data [52], which was sequenced with the reference genome of *S. pohuashanensis*.

*4.6. Metabolomic Analysis*

Using vacuum freeze-drying technology, the embryos of *S. pohuashanensis* were placed in a freeze-drying machine (Scientz-100 F, Scientz Group, Ningbo, China), and the samples were crushed into powder by (MM 400, Retsch, Haan, Germany) type pulverizer. Then, 50 mg of the sample powder was weighed using an electronic balance (MS105DM), and 1200 μL of −20 °C pre-cooled 70% methanolic aqueous internal standard extract was added (less than 50 mg at a rate of 1200 μL extractant per 50 mg sample). It was shaken once every 30 min for 30 s, for a total of 6 times. The treated samples were centrifuged and filtered for UPLC-MS/MS analysis.

The subsequent data processing was implemented by the statistical functions prcomp, cor, ComplexHeatmap 2.8.0, and MetaboAnalystR 1.0.1 in the software package R 3.5.1 (www.r-project.org).

### 4.7. Quantitative Real-Time PCR

To further verify the reliability of the differential expression results of each treatment group in *S. pohuashanensis* samples, we randomly selected 10 genes for qRT-PCR to verify gene expression changes. Using the actin gene of *S. pohuashanensis* as an internal reference, 13 target gene (Table 2 for primer information) primers were designed using Primer 6 software, and the relative expression was expressed by $2^{-\Delta\Delta Ct}$.

**Table 2.** qRT-PCR primers for candidate genes.

| Gene_id | Forward Primers | Reverse Primers |
|---|---|---|
| Spβ-actin | TGGATGGCTGGAAGAGGA | GAGCGGGAAATTGTGAGG |
| Sp01G012390 | GGAAAGCAGCCGAAATCT | TCTCCCACACACCGAAGT |
| Sp02G002460 | GATGTGTATTGCCCTCCT | TCTCTTTACCACCGTTGA |
| Sp02G017590 | GGGTCGGGTTACAAGCG | GGGCAAGAAAACTCGGTG |
| Sp02G022750 | AGGGTCCTCCCCCATACA | CGCTCTCCTTAGCAGTCA |
| Sp03G004740 | CAACCATCCCAAGCCACA | AAACTCCTCCCCTCCTCC |
| Sp03G012770 | CCGTTTATTCCGCCATCA | GGCGTCCCTTTCTTCTCTT |
| Sp04G014660 | GCACGGAGAGGCTAGAGA | GTTGCAGCGAACAATGGT |
| Sp05G009470 | CTGTTCCATTCCTATCCC | CAAAACCCAGTACCCATC |
| Sp13G007860 | TTCTCCGTGTTCGTCCAA | ACCGCCTCCTGTCATTCC |
| Sp14G022200 | GCGGCGAACAAATAGGGT | TGGCGGATAAGGAACGGA |

### 4.8. Statistical Analyses

Data from three biological replicates were analyzed using GraphPad Prism 9, Excel 2019, and SPSS 20.0 and are expressed as mean ± SD. Using one-way ANOVA, $p < 0.05$ was considered significant.

## 5. Conclusions

This study showed that exogenous NO concentrations affect embryo dormancy and germination of *S. pohuashanensis* in a biphasic concentration-dependent effect. In seed germination and seedling morphogenesis, appropriate NO concentrations promoted seed germination by increasing NO and $H_2O_2$ content, reducing ABA content, and increasing ethylene and JA content in *S. pohuashanensis* embryos. However, high concentrations of NO can increase the content of NO in *S. pohuashanensis* embryos. However, high concentrations of NO inhibited the activity of antioxidant enzymes in the embryo of *S. pohuashanensis*, disrupted the hormone balance of the embryo of *S. pohuashanensis*, and inhibited seed germination. A comprehensive multi-omics analysis revealed that NO concentration-dependently regulated *S. pohuashanensis* embryo germination, which was mainly regulated by biological pathways such as phenylpropanoid and flavonoid biosynthesis. Moreover, the biological events that occurred at the transcriptional level were consistent with the phenotype, hormone characteristics, and qRT-PCR analysis. This study provides the major target genes and metabolites of NO in regulating embryo germination of *S. pohuashanensis* and provides an important basis for the molecular mechanism of NO in seed germination of other Rosaceae plants.

**Supplementary Materials:** The following supporting information can be downloaded at: https://www.mdpi.com/article/10.3390/plants14030344/s1, Figure S1. KEGG analysis of DEGs during NO-dependent concentration regulation of embryo germination of Sorbus pohuashanensis. (a) KEGG enrichment statistics of PG-3h vs. CK-3h in non-overlapping DEGs of PG-3h vs. CK-3h and IG-3h vs. CK-3h. (b) KEGG enrichment statistics of IG-3h vs. CK-3h in non-overlapping DEGs of PG-3h vs. CK-3h and IG-3h vs. CK-3h. (c) KEGG enrichment statistics of PG-3h vs. CK-3h in non-overlapping DEGs of PG-3h vs. CK-3h and PG-3d vs. CK-3d. (d) KEGG enrichment statistics of PG-3d vs.

CK-3d in non-overlapping DEGs of PG-3h vs. CK-3h and PG-3d vs. CK-3d. (e) KEGG enrichment statistics of IG-3h vs. CK-3h in IG-3h vs. CK-3h and IG-3d vs. CK-3d non-overlapping DEGs. (f) KEGG enrichment statistics of IG-3d vs. CK-3d in non-overlapping DEGs of IG-3h vs. CK-3h and IG-3d vs. CK-3d. (g) KEGG enrichment statistics of PG-3d vs. CK-3d in non-overlapping DEGs of PG-3d vs. CK-3d and IG-3d vs. CK-3d. (h) KEGG enrichment statistics of IG-3d vs. CK-3d in non-overlapping DEGs of PG-3d vs. CK-3d and IG-3d vs. CK-3d. Figure S2. GO analysis of DEGs during NO-dependent concentration regulation of embryo germination of *Sorbus pohuashanensis*. Figure S3. KEGG enrichment analysis of metabonomics in NO-dependent concentration regulation of *Sorbus pohuashanensis* embryo process.

**Author Contributions:** L.Y. made the research plan, C.Z. wrote the main manuscript text and prepared Figures, C.Z. and Y.Z. review and editing. L.Y. revised the manuscript. All authors have read and agreed to the published version of the manuscript.

**Funding:** This work is supported by the National Natural Science Foundation of China (Grant No. 32071757).

**Data Availability Statement:** The datasets supporting the conclusions of this article are included within the article and its Supplementary Files. All RNA-seq reads were deposited at NCBI (PRJNA1153935).

**Acknowledgments:** We thank the editors and reviewers for their careful reading and valuable comments. We also thank Kombi Kaviriri David of Northeast Forestry University for proofreading.

**Conflicts of Interest:** The authors declare no conflicts of interest.

## References

1. Yang, J.; Li, Y.; Li, J.; Yuan, J.; Wang, S.; Zhou, L.; Zhou, L.; Kang, C.; Guo, L. High-throughput Screening of Secondary Metabolites by *Sorbus pohuashanensis* Cells under Environmental Stress using UHPLC-QTOF Combined with AntDAS. *Physiol. Plant.* **2021**, *173*, 2216–2225. [CrossRef] [PubMed]
2. Yang, L.; Liu, C.P.; Shen, H.L. Effect of Cold Stratification Durations and Germination Temperatures on Seed Germination of *Sorbus pohuashanensis*. *Seed* **2008**, *27*, 20–22+25. [CrossRef]
3. Bethke, P.C.; Libourel, I.G.L.; Vitecek, J.; Jones, R.L. Nitric Oxide Methods in Seed Biology in Seed Dormancy. *Biochem. Mol. Biol.* **2011**, *773*, 385–400. [CrossRef]
4. Bethke, P.C.; Gubler, F.; Jacobsen, J.V.; Jones, R.L. Dormancy of *Arabidopsis* seeds and barley grains can be broken by nitric oxide. *Planta* **2004**, *219*, 847–885. [CrossRef]
5. Gniazdowska, A.; Krasuska, U.; Czajkowska, K.; Bogatek, R. Nitric oxide, hydrogen cyanide and ethylene are required in the control of germination and undisturbed development of young apple seedlings. *Plant Growth Regul.* **2010**, *61*, 75–84. [CrossRef]
6. Wang, Y.T.; Zhao, C.H.; Wang, X.D.; Shen, H.L.; Yang, L. Exogenous Ethylene Alleviates the Inhibition of *Sorbus pohuashanensis* Embryo Germination in a Saline-Alkali Environment ($NaHCO_3$). *Int. J. Mol. Sci.* **2023**, *24*, 4244. [CrossRef]
7. Yang, L.; Shen, H.L.; Liang, L.D.; Liu, C.P. Changes in endogenous hormone content in seeds of *Sorbus pohuashanensis* (Hance) Hedl. during artificial drying and cold stratification. *Plant Physiol. Commun.* **2008**, *44*, 682–688.
8. Li, W.; Niu, Y.; Zheng, Y.; Wang, Z. Advances in the Understanding of Reactive Oxygen Species-Dependent Regulation on Seed Dormancy, Germination, and Deterioration in Crops. *Front. Plant Sci.* **2022**, *13*, 826809. [CrossRef]
9. Zhou, Y.B.; Yun, Y.; Su, B.L.; Chen, Z.J.; Liu, L.W. Effects of Exogenous Nitric Oxide Donor on Seed Germination and Seedling Growth of Several Plant Species. *Plant Physiol. Commun.* **2009**, *41*, 316–318.
10. Krasuska, U.; Ciacka, K.; Orzechowski, S.; Fettke, J.; Bogatek, R.; Gniazdowska, A. Modification of the endogenous NO level influences apple embryos dormancy by alterations of nitrated and biotinylated protein patterns. *Planta* **2016**, *244*, 877–891. [CrossRef]
11. Zhang, Y.; Wang, R.; Wang, X.; Zhao, C.; Shen, H.; Yang, L. Nitric Oxide Regulates Seed Germination by Integrating Multiple Signalling Pathways. *Int. J. Mol. Sci.* **2023**, *24*, 9052. [CrossRef] [PubMed]
12. Ciacka, K.; Staszek, P.; Sobczynska, K.; Krasuska, U.; Gniazdowska, A. Nitric Oxide in Seed Biology. *Int. J. Mol. Sci.* **2022**, *23*, 14951. [CrossRef] [PubMed]
13. Trapet, P.; Kulik, A.; Lamotte, O.; Jeandroz, S.; Bourque, S.; Nicolas-Francès, V.; Rosnoblet, C.; Besson-Bard, A.; Wendehenne, D. NO signaling in plant immunity: A tale of messengers. *Phytochemistry* **2015**, *112*, 72–79. [CrossRef] [PubMed]
14. Corpas, F.J.; del Río, L.A.; Palma, J.M. Impact of Nitric Oxide (NO) on the ROS Metabolism of Peroxisomes. *Plants* **2019**, *8*, 37. [CrossRef]

15. Bailly, C. The signalling role of ROS in the regulation of seed germination and dormancy. *Biochem. J.* **2019**, *476*, 3019–3032. [CrossRef]
16. González-Gordo, S.; Bautista, R.; Claros, M.G.; Cañas, A.; Palma, J.M.; Corpas, F.J. Nitric oxide-dependent regulation of sweet pepper fruit ripening. *J. Exp. Bot.* **2019**, *70*, 4557–4570. [CrossRef]
17. Corpas, F.J.; Freschi, L.; Rodríguez-Ruiz, M.; Mioto, P.T.; González-Gordo, S.; Palma, J.M. Nitro-oxidative metabolism during fruit ripening. *J. Exp. Bot.* **2018**, *69*, 3449–3463. [CrossRef]
18. Rodríguez-Ruiz, M.; González-Gordo, S.; Cañas, A.; Campos, M.J.; Paradela, A.; Corpas, F.J.; Palma, J.M. Sweet Pepper (*Capsicum annuum* L.) Fruits Contain an Atypical Peroxisomal Catalase That Is Modulated by Reactive Oxygen and Nitrogen Species. *Antioxidants* **2019**, *8*, 374. [CrossRef]
19. Corpas, F.J.; González-Gordo, S.; Palma, J.M. Protein nitration: A connecting bridge between nitric oxide (NO) and plant stress. *Plant Stress* **2021**, *2*, 100026. [CrossRef]
20. Jiang, J.L.; Ren, X.M.; Li, L.; Hou, R.P.; Sun, W.; Jiao, C.J.; Yang, N.; Dong, Y.X. $H_2S$ Regulation of Metabolism in Cucumber in Response to Salt-Stress Through Transcriptome and Proteome Analysis. *Front. Plant Sci.* **2020**, *11*, 1283. [CrossRef]
21. Goyal, V.; Jhanghel, D.; Mehrotra, S. Emerging Warriors against Salinity in Plants—Nitric Oxide and Hydrogen Sulphide. *Physiol. Plant.* **2021**, *171*, 896–908. [CrossRef] [PubMed]
22. Arc, E.; Galland, M.; Godin, B.; Cueff, G.; Rajjou, L. Nitric oxide implication in the control of seed dormancy and germination. *Front. Plant Sci.* **2013**, *4*, 346. [CrossRef] [PubMed]
23. Libourel, I.G.L.; Bethke, P.C.; De Michele, R.; Jones, R.L. Nitric oxide gas stimulates germination of dormant *Arabidopsis* seeds: Use of a flow-through apparatus for delivery of nitric oxide. *Planta* **2006**, *223*, 813–820. [CrossRef] [PubMed]
24. Andryka-Dudek, P.; Ciacka, K.; Wiśniewska, A.; Bogatek, R.; Gniazdowska, A. Nitric Oxide-Induced Dormancy Removal of Apple Embryos Is Linked to Alterations in Expression of Genes Encoding ABA and JA Biosynthetic or Transduction Pathways and RNA Nitration. *Int. J. Mol. Sci.* **2019**, *20*, 1007. [CrossRef] [PubMed]
25. Liu, J.; Xue, T.; Shen, Y. Effect of Nitric Oxide on Seed Germination and Dormancy in Empress Trees. *HortTechnology Hortte* **2019**, *29*, 271–275. [CrossRef]
26. Hussain, S.; Nisar, F.; Bhatt, A.; Nielsen, B.L.; Gul, B.; Hameed, A. Redox priming alleviates dormancy and improves salinity tolerance of seeds and seedlings of medicinal halophyte *Zygophyllum simplex* L. *J. Appl. Res. Med. Aromat. Plants* **2022**, *30*, 100384. [CrossRef]
27. Gupta, K.J.; Fernie, A.R.; Kaiser, W.M.; van Dongen, J.T. On the origins of nitric oxide. *Trends Plant Sci.* **2011**, *16*, 160–168. [CrossRef]
28. Ciacka, K.; Krasuska, U.; Otulak-Kozieł, K.; Gniazdowska, A. Dormancy removal by cold stratification increases glutathione and *S*-nitrosoglutathione content in apple seeds. *Plant Physiol. Biochem.* **2019**, *138*, 112–120. [CrossRef]
29. Lopes-Oliveira, P.J.; Oliveira, H.C.; Kolbert, Z.; Freschi, L. The light and dark sides of nitric oxide: Multifaceted roles of NO in plant responses to light. *J. Exp. Bot.* **2020**, *72*, 885–903. [CrossRef]
30. Jahnová, J.; Luhová, L.; Petřivalský, M. S-Nitrosoglutathione Reductase—The Master Regulator of Protein S-Nitrosation in Plant NO Signaling. *Plants* **2019**, *8*, 48. [CrossRef]
31. Albertos, P.; Tatematsu, K.; Mateos, I.; Sánchez-Vicente, I.; Fernández-Arbaizar, A.; Nakabayashi, K.; Nambara, E.; Godoy, M.; Franco, J.M.; Solano, R.; et al. Redox feedback regulation of *ANAC089* signaling alters seed germination and stress response. *Cell Rep.* **2021**, *35*, 109263. [CrossRef] [PubMed]
32. Artur, M.A.S.; Faria, J.M.R.; José, A.C.; Lira, J.M.S.; Souza, K.R.D.; Nery, F.C.; Alvarenga, A.A. Dormancy breaking and biochemical processes associated with germination of *Erythrina falcata* Benth. seeds. *CERNE* **2023**, *29*, e-103170. [CrossRef]
33. Sehrawat, A.; Sougrakpam, Y.; Deswal, R. Cold modulated nuclear S-nitrosoproteome analysis indicates redox modulation of novel Brassicaceae specific, myrosinase and napin in *Brassica juncea*. *Environ. Exp. Bot.* **2019**, *161*, 312–333. [CrossRef]
34. Chaki, M.; De Morales, P.Á.; Ruiz, C.; Begara-Morales, J.C.; Barroso, J.B.; Corpas, F.J.; Palma, J.M. Ripening of pepper (*Capsicum annuum*) fruit is characterized by an enhancement of protein tyrosine nitration. *Ann. Bot.* **2015**, *116*, 637–647. [CrossRef]
35. Wang, H.; Tang, S.; Wang, J.; Shen, H.; Yang, L. Interaction between reactive oxygen species and hormones during the breaking of embryo dormancy in *Sorbus pohuashanensis* by exogenous nitric oxide. *J. For. Res.* **2022**, *33*, 435–444. [CrossRef]
36. Clark, D.; Durner, J.; Navarre, D.A.; Klessig, D.F. Nitric Oxide Inhibition of Tobacco Catalase and Ascorbate Peroxidase. *Mol. Plant-Microbe Interact.* **2000**, *13*, 1380–1384. [CrossRef]
37. Li, J.; Liu, Y.; Zhang, M.; Xu, H.; Ning, K.; Wang, B.; Chen, M. Melatonin increases growth and salt tolerance of *Limonium bicolor* by improving photosynthetic and antioxidant capacity. *BMC Plant Biol.* **2022**, *22*, 16. [CrossRef]
38. Srivastava, A.K.; Singh, D. Assessment of malathion toxicity on cytophysiological activity, DNA damage and antioxidant enzymes in root of *Allium cepa* model. *Sci. Rep.* **2020**, *10*, 886. [CrossRef]
39. Hussain, Q.; Zheng, M.; Hänninen, H.; Bhalerao, R.P.; Riaz, M.W.; Sajjad, M.; Zhang, R.; Wu, J. Effect of the photoperiod on bud dormancy in *Liriodendron chinense*. *J. Plant Physiol.* **2022**, *279*, 153835. [CrossRef]

40. Bykova, N.V.; Hu, J.; Ma, Z.; Igamberdiev, A.U. The Role of Reactive Oxygen and Nitrogen Species in Bioenergetics, Metabolism, and Signaling During Seed Germination. *Agrofor. Syst.* **2015**, *23*, 177–195. [CrossRef]
41. Li, W.; Li, W.F.; Yang, S.J.; Ma, Z.H.; Zhou, Q.; Mao, J.; Han, S.Y.; Chen, B. Transcriptome and Metabolite Conjoint Analysis Reveals that Exogenous Methyl Jasmonate Regulates Monoterpene Synthesis in Grape Berry Skin. *J. Agric. Food Chem.* **2020**, *68*, 5270–5281. [CrossRef] [PubMed]
42. Rahman, M.M.; Mostofa, M.G.; Keya, S.S.; Ghosh, P.K.; Abdelrahman, M.; Anik, T.R.; Gupta, A.; Tran, L.P. Jasmonic acid priming augments antioxidant defense and photosynthesis in soybean to alleviate combined heat and drought stress effects. *Plant Physiol. Biochem.* **2024**, *206*, 108193. [CrossRef] [PubMed]
43. Xing, J.; Cao, X.; Zhang, M.; Wei, X.; Zhang, J.; Wan, X. Plant nitrogen availability and crosstalk with phytohormones signallings and their biotechnology breeding application in crops. *Plant Biotechnol. J.* **2023**, *21*, 1320–1342. [CrossRef]
44. Tong, Y.; Liu, S.Y.; Yi, S.C.; Qiu, Z.X.; Wang, Y.H.; Zeng, D.Q.; Tang, W.W. Bruceine D, the main active ingredient of *Brucea javanica* (L.) residue inhibits the germination of *Bidens pilosa* L. seeds by suppressing phenylpropanoid biosynthesis. *Ind. Crops Prod.* **2021**, *172*, 114079. [CrossRef]
45. Agati, G.; Brunetti, C.; Di Ferdinando, M.; Ferrini, F.; Pollastri, S.; Tattini, M. Functional roles of flavonoids in photoprotection: New evidence, lessons from the past. *Plant Physiol. Biochem.* **2013**, *72*, 35–45. [CrossRef]
46. Chen, X.T.; Zhao, G.Y.; Li, Y.L.; Wei, S.M.; Dong, Y.H.; Jiao, R.Z. Integrative Analysis of the Transcriptome and Metabolome Reveals the Mechanism of Chinese Fir Seed Germination. *Forests* **2023**, *14*, 676. [CrossRef]
47. Liu, H.; Chen, X.X.; Chen, H.X.; Lu, J.; Chen, D.L.; Luo, C.; Cheng, X.; Jia, Y.; Huang, C.L. Transcriptome and Metabolome Analyses of the Flowers and Leaves of *Chrysanthemum dichrum*. *Front. Genet.* **2021**, *12*, 716163. [CrossRef]
48. Ahmed, U.; Rao, M.J.; Qi, C.; Xie, Q.; Noushahi, H.A.; Yaseen, M.; Shi, X.; Zheng, B. Expression Profiling of Flavonoid Biosynthesis Genes and Secondary Metabolites Accumulation in *Populus* under Drought Stress. *Molecules* **2021**, *26*, 5546. [CrossRef]
49. Gao, Q.; Li, X.; Xiang, C.; Li, R.; Xie, H.; Liu, J.; Li, X.; Zhang, G.; Yang, S.; Liang, Y.; et al. *EbbHLH80* Enhances Salt Responses by Up-Regulating Flavonoid Accumulation and Modulating ROS Levels. *Int. J. Mol. Sci.* **2023**, *24*, 11080. [CrossRef]
50. Yang, L.; Shen, H.L. Effect of electrostatic field on seed germination and seedling growth of *Sorbus pohuashanesis*. *J. For. Res.* **2011**, *22*, 27–34. [CrossRef]
51. Wang, X.D.; Shen, H.L.; Yang, L. The Response of Hormones, Reactive Oxygen Species and Nitric Oxide in the Polyethylene-Glycol-Promoted, Salt-Alkali-Stress-Induced Embryo Germination of *Sorbus pohuashanensis*. *Int. J. Mol. Sci.* **2024**, *25*, 5128. [CrossRef] [PubMed]
52. Zhao, D.; Zhang, Y.; Lu, Y.; Fan, L.; Zhang, Z.; Zheng, J.; Chai, M. Genome sequence and transcriptome of *Sorbus pohuashanensis* provide insights into population evolution and leaf sunburn response. *J. Genet. Genom.* **2022**, *49*, 547–558. [CrossRef]

*plants*

*Article*

# The Establishment of a Highly Efficient In Vitro Regeneration System for *Viburnum opulus* L. 'Roseum'

**Yajing Ning [1,†], Hao Dong [1,†], Xinxin Zhang [1], Yanhua Li [2], Chengpeng Cui [1] and Shujuan Li [1,*]**

1 State Key Laboratory of Tree Genetics and Breeding, Northeast Forestry University, Harbin 150040, China
2 Zhejiang Physical and Chemical Technology Co., Ltd., Hangzhou 310000, China
* Correspondence: lishujuan@nefu.edu.cn
† These authors contributed equally to this work.

**Abstract**

*Viburnum opulus* L. 'Roseum' is a highly valuable ornamental plant for landscaping, but it has a long propagation cycle and low propagation coefficient. In this study, stem segments with axillary buds from *Viburnum opulus* L. 'Roseum' were used as explants. We systematically analyzed the use of sodium hypochlorite for the sterilization of explants, as well as the effects of different plant growth regulator combinations and concentrations on shoot bud induction, shoot proliferation, the rooting of tissue-cultured shoots, and the transplanting of the tissue-cultured shoots. A complete rapid propagation technology system for *Viburnum opulus* L. 'Roseum' was established. The results showed that a disinfection method using 75% ethanol for 30 s and soaking in 5% sodium hypochlorite for 5 min was the most suitable for disinfecting the stem segments of *Viburnum opulus* L. 'Roseum', which showed low contamination and a 73.33% survival rate. The ideal medium for primary bud induction was WPM (Woody Plant Basal Medium) + 2.0 mg·L$^{-1}$ 6-benzylaminopurine (6-BA) + 0.15 mg·L$^{-1}$ indole-3-butyric acid solution (IBA) + 25 g·L$^{-1}$ sucrose. The optimal medium for shoot proliferation was WPM + 1.0 mg·L$^{-1}$ 6-BA + 0.15 mg·L$^{-1}$ IBA + 25 g·L$^{-1}$ sucrose, achieving an induction rate of 7.17. For the rooting of tissue-cultured shoots, the most suitable formulation was 1/2 WPM + 0.3 mg·L$^{-1}$ naphthaleneacetic acid (NAA) + 0.3 mg·L$^{-1}$ activated charcoal (AC) + 25 g·L$^{-1}$ sucrose, which induced robust and developed root systems. This study provides a technical basis for the establishment of a fast propagation system for the industrial production of *Viburnum opulus* L. 'Roseum'.

**Keywords:** axillary shoot proliferation; micropropagation; plant tissue culture; ornamental plants

## 1. Introduction

The *Viburnum* plant holds a very important position in ornamental plant resources due to its strong adaptability and high ornamental value, earning the title of a 'universal greening tree' [1]. *Viburnum opulus* L. 'Roseum' is a deciduous shrub of *Viburnum* [2]. It is cultivated in China in Jiangsu, Zhejiang, Jiangxi, and Hebei Provinces. The young branches of the year are ribbed and hairless on the surface. The winter buds are oval-shaped and have one pair of fused outer scales. The leaves are round-ovate, broad-ovate, or reverse-ovate, measuring 6–12 cm in length, with leaf lobes, where the middle lobes are long, palmately three-veined, and the base is round, flat, or shallowly heart-shaped, with a hairless surface. The petioles are relatively thick, measuring 1–2 cm in length, and have 2–4 or more disk-shaped glands. The compound umbel inflorescences are 5–10 cm in diameter, featuring large sterile flowers, with thick total peduncles measuring 2–5 cm in length and a hairless surface, and the flower stalks are extremely short. The flower corolla is white and radially symmetrical. The sterile flowers are also white, measuring 1.3–2.5 cm in diameter, and have long stalks, with broad and reverse-ovate lobes. The leaves of *Viburnum opulus* L. 'Roseum' fall in winter and sprout in early spring, remaining green until early autumn. During the transition from autumn to winter, the leaf color changes to red, making it a highly ornamental landscape plant. As an ecological landscaping tree species, *Viburnum opulus* L. 'Roseum' can be planted singly, in clusters, or mixed with other plants. It is suitable for use as a garden tree, street tree, or landscape tree and is also appropriate for large-scale planting as an ornamental tree in botanical gardens and natural scenic areas [2]. *Viburnum opulus* L. 'Roseum' seedlings are difficult to source, and there are many sterile flowers, and cutting propagation suffers from low rooting rates. Therefore, it is of great significance to study the efficient propagation of *Viburnum opulus* L. 'Roseum' to provide technical support for urban greening.

*Viburnum opulus* L. 'Roseum' is an important ornamental plant resource with strong adaptability, making it popular in landscape design. Research by Du Xiao [3] and Bi Bo [4] showed that *Viburnum tinus* and *Viburnum punctatum* have strong heavy metal accumulation capabilities. In addition to its high esthetic value, *Viburnum opulus* L. 'Roseum' also contributes to environmental beautification and has ecological benefits. Currently, the main propagation methods for *Viburnum opulus* L. 'Roseum' are cuttings and stem segments, which have a low propagation capacity. It is challenging to obtain a large number of healthy and disease-free plantlets in a short period due to the long cycle and low propagation coefficient of these methods, resulting in current production not meeting market demand. In recent years, there have been related reports on the propagation technology of *Viburnum opulus.* L. Chen Bihua [5] conducted experiments on cuttings using *Viburnum macrocephalum* Fort. as the material. The results showed that treatment with 500 ppm IBA for 5 min resulted in a rooting rate of 66.5% to 66.9% after 60 days. Bi Xianyu [6] tested cuttings of current year branches of *Viburnum Sargenti* Koehne, and the results surfaced that their survival rates ranged from 53.28% to 61.53%. Plant tissue culture can meet the demands of large-scale plant propagation and serves as an important tool for studying plant resistance to stress [7]. Virginia Hildebrandt [8] used *Viburnum opulus* 'Nanum' as a material and indicated that each bud could produce approximately seven explants after 6–8 months. This indicates that tissue culture is an important method for the rapid propagation of plants, allowing for both efficient reproduction and the preservation of excellent traits in clonal lines.

*Viburnum opulus* L. 'Roseum' has significant landscape ecological value and high market demand. Its propagation coefficient is low and its cycle is relatively long when propagated via cuttings, whereas tissue culture can yield numerous tissue-cultured plantlets in a short time. Currently, reports on the tissue culture techniques for *Viburnum opulus* 'Roseum' indicate a low proliferation rate. Suaad A. Yaseen [9] conducted research using *Viburnum opulus* 'Roseum' as the material, finding a proliferation rate of 3.31. Therefore, we aimed to establish a more efficient rapid propagation tissue culture system for robust *Viburnum opulus* 'Roseum' to provide an effective pathway for large-scale tissue-cultured plantlet production while better meeting the needs of production and the market.

## 2. Results

### 2.1. Surface Sterilization of Explants

The disinfection results of *Viburnum opulus* L. 'Roseum' stem segments treated with different concentrations of sodium hypochlorite and varying disinfection times are shown in Table 1. When a 2% sodium hypochlorite solution was used for disinfection for 5 and

10 min, the stem segments did not die but the contamination rates reached as high as 80.00% to 62.22%, with the highest survival rate being only 37.78%. In contrast, when a 5% sodium hypochlorite solution was used for disinfection for 5 min, the contamination rate decreased by 60% compared with the 2% sodium hypochlorite solution. The highest survival rate at this time was 73.33%. When the same concentration of disinfectant was used for 10 min, a mortality rate of 10.00% was observed, and the survival rate began to decline. After treatment with a 10% sodium hypochlorite solution for 5 min, the contamination rate was only 13.33%, with a mortality rate of over 50%. When the 10% sodium hypochlorite solution was applied for 10 min, the contamination rate dropped to 0.00%, but the mortality rate surged to 75.56%.

**Table 1.** Effect of different treatment combinations on the sterilization of explants.

| Group | Sodium Hypochlorite Concentration | Sodium Hypochlorite Treatment Time | Contamination Rate % | Mortality Rate % | Recovery Rate% |
|---|---|---|---|---|---|
| 1 | 2% | 5 min | 80.00 ± 1.92 a | 0.00 | 20.00 ± 3.33 d |
| 2 | 2% | 10 min | 62.22 ± 1.11 b | 0.00 | 37.78 ± 2.94 b |
| 3 | 5% | 5 min | 24.44 ± 2.94 c | 2.22 ± 1.92 d | 73.33 ± 1.92 a |
| 4 | 5% | 10 min | 21.11 ± 2.22 c | 10.00 ± 1.92 c | 67.78 ± 2.22 a |
| 5 | 10% | 5 min | 13.33 ± 1.92 d | 53.33 ± 6.67 b | 28.89 ± 1.11 c |
| 6 | 10% | 10 min | 0.00 | 75.56 ± 3.85 a | 24.44 ± 1.11 cd |

Note: Different lowercase letters in the same column indicate significant differences between treatments ($p < 0.05$).

## 2.2. Primary Bud Induction Culture

The stem segments that had been disinfected via explant treatment were placed in a culture medium for the initial induction of axillary buds. The germination time of the axillary buds and the state of the induced axillary buds vary among different combinations of plant growth regulators, as shown in Figure 1 and Table 2. In the initial induction Scheme A, axillary buds began to grow after 8.67 days of inoculation, whereas in Scheme B, they only started to grow after 10.33 days. After 45 days, the leaf growth rate of the axillary buds in Scheme A was fast, but the leaves were elongated, slightly curled, and had deeper leaf lobes, with the apical bud not prominent and consisting of several deformed buds. In Scheme B, the growth of the axillary buds from the stem segments was slower, the leaves were flat, and the leaf color was a vibrant green, which indicated normal axillary buds.

**Figure 1.** Growth status of axillary buds of the 2 initial induction schemes. (**a**) Scheme A; (**b**) Scheme B.

**Table 2.** Time to sprouting of axillary buds and axillary bud status after 45 days for the two initial induction protocols.

| Group | Number of Inoculated Stem Segments/pc | Axillary Bud Sprouting Time/Day | Status of Axillary Buds |
|---|---|---|---|
| A | 30 | 8.67 | Leaf blade curled, apical dominance not obvious, more deformed buds |
| B | 30 | 10.33 | Axillary buds grow well, with spreading, bright green leaves |

### 2.3. *Shoot Proliferation*

Table 3 and Figure 2 show that among the nine propagation schemes designed using orthogonal experiments, Scheme 3 had the highest number of axillary shoot buds (7.17), with normal growth, flat leaves, and typical leaf shapes. In contrast, Schemes 2, 6, and 8 had fewer numbers of axillary bud proliferations. In Scheme 2, the differentiated buds were generally reddish and mostly deformed. Scheme 6 produced more deformed buds, characterized by elongated leaves, deeper leaf fissures, and longer petioles. In Scheme 8, the number of axillary buds was low, with curled leaves and short internodes. Schemes 1, 4, 5, 7, and 9 produced 4.17 to 6.33 proliferated buds, but many of these exhibited reddish leaves. In Scheme 4, there was the growth of callus tissue on the stem. In Scheme 5, the proliferated buds had elongated leaves with slightly increased internodes. In Scheme 9, the buds showed poor growth, with yellowing leaves and some basal leaves dying.

**Table 3.** Effect of different concentrations of 6-BA, IBA, and sucrose on adventitious bud differentiation.

| Groups | 6-BA Concentration (mg L$^{-1}$) | IBA Concentration (mg L$^{-1}$) | Sucrose Concentration (g·L$^{-1}$) | Multiplication Coefficient | Growth State of Tissue-Culture Plantlets |
|---|---|---|---|---|---|
| 1 | 1.0 | 0.05 | 20 | 5.83 ± 0.31 abc | Leaves yellowing, differentiation buds turning red, growth is average. |
| 2 | 1.0 | 0.1 | 30 | 3.67 ± 0.42 e | Few differentiation buds, signs of vitrification, stems and leaves are thick and red. |
| 3 | 1.0 | 0.15 | 25 | 7.17 ± 0.48 a | More differentiation buds, leaf shape normal, growth is good. |
| 4 | 1.5 | 0.05 | 30 | 4.17 ± 0.31 de | Growth is average, stems have hyperplastic tissue proliferation. |
| 5 | 1.5 | 0.1 | 25 | 4.67 ± 0.49 cde | Growth is average, leaves are slender, leaf edges turning red. |
| 6 | 1.5 | 0.15 | 20 | 3.83 ± 0.31 de | Leaves are slender, most differentiation buds are malformed. |
| 7 | 2.0 | 0.05 | 25 | 5.17 ± 0.31 bcd | Growth is average, leaf edges slightly red, and some malformed buds present. |
| 8 | 2.0 | 0.1 | 20 | 3.50 ± 1.0.43 e | Growth is average, fewer differentiation buds, leaves curling. |
| 9 | 2.0 | 0.15 | 30 | 6.33 ± 0.766 ab | Growth is poor, leaves yellowing, more differentiation buds. |

Note: Different lowercase letters in the same column indicate significant differences between treatments ($p < 0.05$).

### 2.4. *Induction of Adventitious Roots*

During the experiment, the tissue-cultured plantlets without added activated charcoal (AC) had less rooting, the main root was thicker, there was a whitish epidermis of the main root, and slow growth, as shown in Figure 3. However, adding an appropriate amount of AC improved the growth environment of the tissue-cultured plantlets in the medium, leading to healthier root growth. Therefore, the rooting culture phase for *Viburnum opulus* L. 'Roseum' tissue-cultured plantlets must include an appropriate amount of AC in the medium.

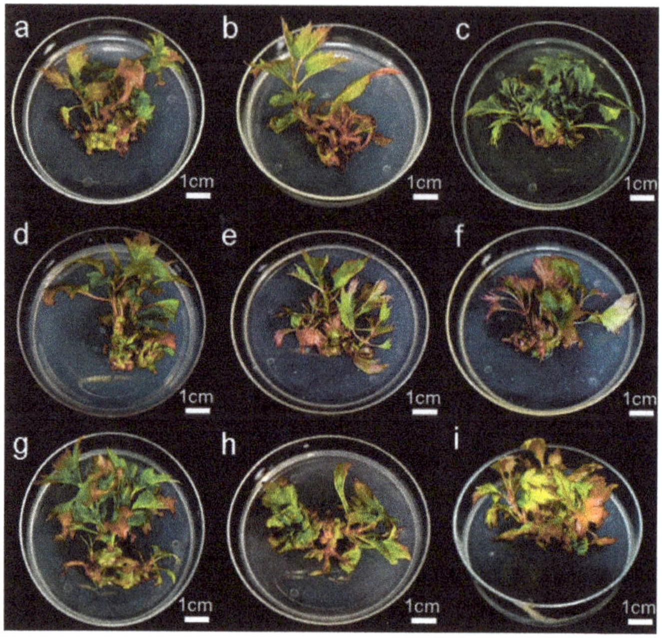


**Figure 2.** Effect of different concentrations of 6-BA, IBA, and sucrose on axillary bud differentiation. (**a–i**) are groups 1–9, respectively.

**Figure 3.** Effect of not adding AC to the medium on rooting. Both (**a**,**b**) show the state of the root system of the tissue-culture plantlets when no AC was added.

As shown in Table 4 and Figure 4, in the nine different rooting treatments, there was still swelling in the epidermal layer of the primary roots with an AC concentration of 0.1 g·L$^{-1}$. When the rooting regimen was 1/2 WPM + 0.3 mg·L$^{-1}$ NAA + 0.3 mg·L$^{-1}$ AC + 25 g·L$^{-1}$ sucrose, the tissue-cultured plantlets had a high rooting rate, relatively fast growth rate, and vigorous root system. And the root system was robust, the plants grew normally, and the leaves were flat without curling. However, as the concentration of NAA gradually increased, the rooting rate of *Viburnum opulus* L. 'Roseum' was inhibited, leading to slower root growth. The rooting rate of Group 3 was as high as 88.89%, followed by those of Groups 4 and 5, with a difference of only 11.11%. However, the adventitious roots of the tissue-cultured plantlets in Group 3 were robust and had a higher mean number of primary roots of 4.38, while the mean number of primary roots in Group 4 was 4.29.

**Table 4.** Effect of different concentrations of sucrose, NAA, and AC on rooting.

| Groups | Sucrose Concentration (g·L$^{-1}$) | NAA Concentration (mg L$^{-1}$) | AC Concentration (mg L$^{-1}$) | Rooting Rate % | Average Number of Primary Roots | Tissue-Culture Plantlets Grow and Root |
|---|---|---|---|---|---|---|
| 1 | 25 | 0.1 | 0.1 | 23.46 ± 1.23 d | 2.00 ± 50.58 d | Tissue-cultured plantlets grew poorly, browning at the base with only a few roots. |
| 2 | 25 | 0.2 | 0.5 | 56.79 ± 3.27 c | 2.80 ± 0.17 bc | Tissue-cultured plantlets have average growth with few roots. |
| 3 | 25 | 0.3 | 0.3 | 88.89 ± 3.70 a | 4.38 ± 0.08 a | Tissue-cultured plantlets have good growth with strong primary roots. |
| 4 | 30 | 0.1 | 0.5 | 77.78 ± 4.28 b | 4.29 ± 0.10 a | Tissue-cultured plantlets with good growth, weak primary root. |
| 5 | 30 | 0.2 | 0.3 | 77.78 ± 3.70 b | 2.29 ± 0.05 cd | Tissue culture plantlets were average in growth, with late root formation and short primary roots. |
| 6 | 30 | 0.3 | 0.1 | 55.56 ± 4.28 c | 3.40 ± 0.21 b | Tissue-culture plantlets have poor growth, with few main roots, thickened epidermis, and the growth of callus at the base. |
| 7 | 35 | 0.1 | 0.3 | 55.56 ± 2.14 c | 3.40 ± 0.10 b | Tissue-culture plantlets have average growth, with few roots and a small number of secondary roots. |
| 8 | 35 | 0.2 | 0.1 | 56.79 ± 3.27 c | 2.20 ± 0.07 cd | Tissue-culture plantlets have average growth, with a thickened epidermis on the main root. |
| 9 | 35 | 0.3 | 0.5 | 55.56 ± 3.70 c | 1.20 ± 0.05 e | Tissue-culture plantlets have poor growth, with weak and short main roots and browning at the base. |

Note: Different lowercase letters in the same column indicate significant differences between treatments ($p < 0.05$).

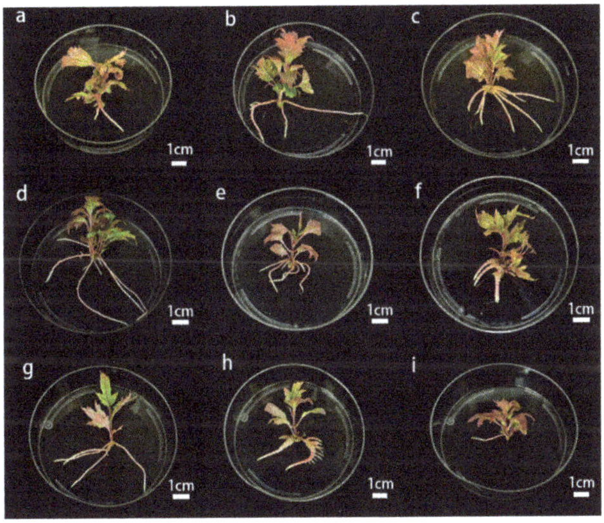


**Figure 4.** Effect of different concentrations of sucrose, NAA, and AC on rooting. (**a**–**i**) are groups 1–9, respectively.

## 2.5. Transplantation

The rooting culture of *Viburnum opulus* L. 'Roseum' tissue-cultured plantlets was carried out based on the rooting protocol of Group 3. The tissue-cultured plantlets of *Viburnum opulus* L. 'Roseum' adapted to the transplanting environment after indoor acclimatization, resulting in the successful survival of all plantlets post-transplant. As shown in Figure 5, different substrates had varying effects on the transplantation of *Viburnum opulus* L. 'Roseum' tissue-cultured plantlets. When the tissue-cultured plantlets were cultivated in a substrate mix of peat soil and vermiculite in a ratio of 1:1 for 10 days, the surviving plantlets exhibited curled leaves that were reddish-brown, with stunted development and overall smaller stature. After 10 days of culture in a substrate mixed in a ratio of 1:1:1 of peat soil, perlite, and vermiculite, some leaves of the surviving small plants were curled, and growth was somewhat slow. When the tissue-cultured plantlets were planted in a substrate mix of peat soil and vermiculite at a ratio of 1:2 for 10 days, some leaves showed reddish-brown coloration, and the growth was somewhat slow. In the substrate mix of peat soil, vermiculite, and perlite at a ratio of 1:2:1, the tissue-cultured plantlets grew vigorously, demonstrating healthy growth and normal leaf extension. After 40 days of growth, the plants were robust, with leaves displaying a shiny appearance, and the young buds at the base also began to grow. Therefore, we preliminarily conclude that the optimal substrate for the growth of *Viburnum opulus* L. 'Roseum' has a ratio of 1:2:1 of peat soil, vermiculite, and perlite. In addition, the survival rate of the tissue-cultured plantlets in different substrate ratios, which only showed differences in terms of growth status, was maintained at 100%.

**Figure 5.** Effects of different substrates on the growth of *Viburnum opulus* L. 'Roseum' tissue-cultured plantlets. (**a**) Peat soil–vermiculite = 1:1; (**b**) peat soil–vermiculite–perlite = 1:1:1; (**c**) peat soil–vermiculite = 1:2; (**d**) peat soil–vermiculite–perlite = 1:2:1; (**e**) after 45 days incubation in a mixed substrate of peat soil–vermiculite–perlite at 1:2:1.

## 3. Discussion

Our current understanding indicates that the propagation efficiency of *Viburnum opulus* L. 'Roseum' is relatively low. Therefore, in this study, we established a system for the efficient in vitro culture of *Viburnum opulus* L. 'Roseum', comprising five stages: explant disinfection, axillary bud initial induction, multiple bud induction, rooting of shoots, and substrate selection. In sterilants, mercuric chloride and sodium hypochlorite are widely used [10]. However, mercuric chloride solution tends to leave residues on the surface of explants, causing significant damage to the explants. If not rinsed thoroughly with sterile water, it can lead to the death of the explants [11]. Therefore, in this experiment, sodium hypochlorite was selected for surface disinfection of the explants. In the experiment, after cutting the explants, they were first rinsed to remove dust and debris for 30 min under running water. Subsequently, disinfecting with 5% sodium hypochlorite for 5 min proved to be the most suitable approach for disinfection, as it not only increased the survival rate of the explants but also reduced contamination rates.

The induction rate of adventitious buds and the propagation coefficient are important indicators during the initial induction and proliferation phases and are influenced by various factors, including the culture medium and plant growth regulator [12]. Jian Defeng [13] found that the combination of KT (kinetin) and IBA plant growth regulator significantly improved propagation effectiveness in *Viburnum sargentii* Koehne. In this experiment, in the initial induction stage of *Viburnum opulus* L. 'Roseum', it was discovered that the medium WPM (Woody Plant Basal Medium) + 0.15 mg$\cdot$L$^{-1}$ indole-3-butyric acid solution (IBA) + 2.0 mg$\cdot$L$^{-1}$ 6-benzylaminopurine (6-BA) + 25 g$\cdot$L$^{-1}$ sucrose was most suitable for initial induction. When the concentration of 6-BA was 2.0 mg$\cdot$L$^{-1}$, the induced axillary bud leaves were deformed, and apical dominance was not evident. Wang Hongbao [14] found that a medium of MS + 0.5 mg$\cdot$L$^{-1}$ 6-BA + 0.2 mg$\cdot$L$^{-1}$ NAA was effective for *Viburnum sargentii*, with the propagation coefficient reaching 7. Wu Ju [15] discovered that the combination of 6-BA and IBA plant growth regulator was better suited for the propagation of adventitious buds in *Viburnum sargentii*, when the proliferation factor increased by up to six times. Zhen Xuehua [16] conducted tissue culture studies using young stems of *Viburnum opulus* L. as explants, based on MS (MA Base Salts) medium, achieving a multiplication rate of 2.95. In this study, it was found that the best propagation scheme used a medium of 0.15 mg·L$^{-1}$ IBA + 1.0 mg·L$^{-1}$ 6-BA + 25 g·L$^{-1}$ sucrose, with a differentiation coefficient of 7.17 for adventitious buds. When the concentration of 6-BA increased, most of the adventitious buds became deformed, with yellowing leaves or overall reddening of the plant. The reasons for the color change in plant leaves are numerous. *Viburnum opulus* L. 'Roseum', as a deciduous tree species that exhibits autumn coloration, is very sensitive to changes in light and temperature. We speculate that during its cultivation, subtle variations in the external environment, as well as the effects of excessively high or low levels of plant growth regulators and carbon sources in the culture medium, lead to the formation of colored buds and malformed buds.

The rooting and acclimatization of tissue-cultured plantlets are key steps for the large-scale and commercial application of tissue culture rapid propagation technology [17]. *Ormosia henryi* Prain [18] and *Vaccinium ashei* 'Tifblue' [19] both utilize 1/2 WPM medium for rooting induction. Du Shurui [20] and others used IAA for the culture rooting of *Viburnum* to achieve a 93.5% rooting rate. Wu Ju [15] utilized IBA to induce rooting in *Viburnum sargentii* tissue-cultured shoots, achieving a rooting rate of 90%. In this experiment using axillary buds to induce rooting, it was found that NAA had a better induction effect, resulting in a more robust and developed root system, which is consistent with the research by Cai Neng and others [21]. The mechanism by which activated carbon promotes root formation is through its role as a physical adsorbent. Activated carbon can enhance the aeration and water-holding capacity of the culture medium, while also absorbing toxic substances secreted by the cultured material, thereby reducing the accumulation of these harmful substances. This leads to reduced oxidative stress in the root system and promotes plant root growth. Luo Mengwei [22] found that activated carbon can significantly improve the rooting rate of tissue-cultured plantlets of *Platostoma palustre*, achieving a rooting rate of 100%. Bi Xianyu [23] and others achieved a rooting rate of 90.91% by adding activated charcoal. In the present study, it was found that the non-addition of an appropriate amount of AC in the rooting medium affected the robustness of the root system of the tissue-cultured shoots to some extent. This is consistent with the findings of Bi Xianyu [6]. Roots that differentiated without AC usually had an enlarged outer epidermis with a whitish color, which was easily damaged by rinsing under running water and could easily cause root composting death of tissue-cultured plantlets during planting. Activated charcoal can provide a suitable dark environment for root development, and the addition of the appropriate amount of activated charcoal can promote the rooting of tissue-cultured plantlets. Therefore, the optimal plant growth regulator concentration for the rooting of axillary buds in *Viburnum opulus* L. 'Roseum' was determined to be 1/2 WPM + 0.3 mg·L$^{-1}$ NAA + 0.3 g·L$^{-1}$ AC + 25 g·L$^{-1}$ sucrose, with rooting beginning at approximately 15 days and the rooting rate reaching 88.89%.

The survival rate of transplanted tissue-cultured plantlets is a key factor in assessing whether the tissue culture technology system is mature. Light, temperature, substrate aeration, water retention, and nutrient capacity are important factors affecting the survival of tissue-cultured plantlets [24,25]. In this experiment, peat soil, vermiculite, and perlite were used as the substrates. The mixed substrate of peat soil, vermiculite, and perlite in a ratio of 1:2:1 was superior to other ratios for growth. The tissue-cultured plantlets of *Viburnum opulus* L. 'Roseum' exhibited deep green leaves and robust growth. This mixed substrate contained abundant moisture and nutrients, meeting the growth demands of *Viburnum opulus* L. 'Roseum' tissue-cultured plantlets.

## 4. Materials and Methods

### 4.1. Plant Materials

Using robust and disease-free *Viburnum opulus* L. 'Roseum' from the laboratory of Northeast Forestry University as the material, new spring branches from the current year were selected as explants to establish a tissue culture rapid propagation system.

### 4.2. Components of the Medium

The details of all the mentioned substances are as follows: Lloyd & McCown Woody Plant Basal Medium with Vitamins (WPM), L449; agar, PLANT TC, A111; 6-benzylaminopurine (6-BA) at a concentration of 1 mg·mL$^{-1}$, T818; indole-3-butyric acid solution (IBA) at 1 mg·mL$^{-1}$; naphthaleneacetic acid (NAA) at 1 mg·mL$^{-1}$; activated carbon (AC), I460, from PhytoTechnology Laboratories; and sucrose, A502792 from Sangon Biotech (Shanghai, China). Peat soil was from Hongli Flower Base, Harbin, Heilongjiang Province, China.

### 4.3. Cultivation Conditions

The WPM medium base consisted of 5.3 g·L$^{-1}$ plant agar, adjusted to a pH of 5.8 ± 0.02. Although the medium formulation remained consistent across various growth stages, the type and concentration of plant growth regulators differed. The medium was sterilized in an autoclave at 121 °C for 20 min. Incubation was maintained at a temperature of 25 ± 1 °C, with a light intensity of 40–50 μmol·m$^{-2}$·s$^{-1}$, following a 16/8 h light–dark cycle.

### 4.4. Sterilization of Explants

Healthy, pest-free, one-year-old fresh branches with bud points were cut and trimmed into small segments approximately 3.0 cm long, each containing a bud point. First, the material was washed in sterile water containing 2.0 mg·L$^{-1}$ Tween 20 for 3–5 min and then covered with gauze and rinsed under running water for 30 min. After rinsing, the material was transferred to a vertical laminar flow cabinet for explant sterilization. The stem segments was washed once with sterile water and then immersed in 75% ethanol for 30 s while shaking thoroughly to ensure the alcohol made sufficient contact with the surface of the stem segments. After this, they were washed twice with sterile water and then disinfected with 2%, 3%, or 5% sodium hypochlorite for 5, 10, or 15 min. After disinfection, they were washed five times with sterile water. The disinfection treatment combinations are listed in Table 1. Finally, sterile filter paper was used to dry the moisture from the surface of the stem segments. Then, sterile scissors was used to trim off the ends that had been oxidized by the sodium hypochlorite. The remaining stem segments should be cut into small segments of about 1.5–2.0 cm and seeded into WPM minimal medium (each stem segment has two axillary buds). Each vial was inoculated with three stem segments and ten vials were replicated for each treatment group for a total of thirty tissue-cultured plantlets. The culture conditions were as follows: illumination for 16 h at an intensity of 40–50 μmol·m$^{-2}$·s$^{-1}$ and a temperature of 25 °C ± 1 °C. After 7 days, the rates of contamination, mortality, and survival of the stem segments were observed and counted as follows: contamination rate (%) = (total number of samples processed/number of contamination samples) × 100%; mortality rate (%) = (total number of treated samples/number of dead samples) × 100%; survival rate (%) = (total number of treated samples/number of surviving samples) × 100%.

### 4.5. Primary Culture of *Viburnum opulus* L. 'Roseum'

The cleaned explants were inoculated onto WPM medium for axillary bud induction. Referring to Bi Xianyu's [23] initial induction protocol, the following initial induction protocol was selected: WPM + 0.2 mg·L$^{-1}$ IBA + 2.0 mg·L$^{-1}$ 6-BA + 25 g·L$^{-1}$ sucrose + 5.3 g·L$^{-1}$ agar (A); WPM + 0.15 mg·L$^{-1}$ IBA + 2.0 mg·L$^{-1}$ 6-BA + 25 g·L$^{-1}$ sucrose + 5.3 g·L$^{-1}$ agar (B). Three explants were inoculated per bottle, with ten replicates for each protocol. The time of sprouting of axillary buds was counted, and the status of the sprouted axillary buds was observed within 45 days.

### 4.6. Induction of Bud Clusters from Young Stems

Using the induced sterile axillary buds as materials, they were transferred to media containing different combinations of plant growth regulators for proliferation culture. An $L_9$ ($3^4$) orthogonal experimental design was used, as shown in Table 5. Three sterile buds were inoculated per bottle, with ten replicates for each treatment. After 30 days of cultivation, the proliferation and growth status of the buds were observed and recorded.

**Table 5.** Table of orthogonal experimental design for axillary bud proliferation program.

| Group | 6-BA Concentration (mg·L$^{-1}$) | IBA Concentration (mg·L$^{-1}$) | Sucrose Concentration (g·L$^{-1}$) |
|---|---|---|---|
| 1 | 1.0 | 0.05 | 20 |
| 2 | 1.0 | 0.1 | 30 |
| 3 | 1.0 | 0.15 | 25 |
| 4 | 1.5 | 0.05 | 30 |
| 5 | 1.5 | 0.1 | 25 |
| 6 | 1.5 | 0.15 | 20 |
| 7 | 2.0 | 0.05 | 25 |
| 8 | 2.0 | 0.1 | 20 |
| 9 | 2.0 | 0.15 | 30 |

### 4.7. Rooting Culture

The robust, sterile tissue-cultured shoots of *Viburnum opulus* L. 'Roseum' were transferred into a rooting culture medium using 1/2 WPM basic nutrient medium. Different concentrations of sucrose, NAA, and AC were designed using an $L_9$ ($3^4$) orthogonal experimental design for the rooting culture, as shown in Table 6. Each treatment was replicated in twelve bottles, with three plants per bottle. The rooting conditions were observed and recorded after 40 days.

**Table 6.** Table of orthogonal experiments for rooting schemes.

| Group | Sucrose Concentration (g·L$^{-1}$) | NAA Concentration (mg·L$^{-1}$) | AC Concentration (mg·L$^{-1}$) |
|---|---|---|---|
| 1 | 25 | 0.1 | 0.1 |
| 2 | 25 | 0.2 | 0.5 |
| 3 | 25 | 0.3 | 0.3 |
| 4 | 30 | 0.1 | 0.5 |
| 5 | 30 | 0.2 | 0.3 |
| 6 | 30 | 0.3 | 0.1 |
| 7 | 35 | 0.1 | 0.3 |
| 8 | 35 | 0.2 | 0.1 |
| 9 | 35 | 0.3 | 0.5 |

### 4.8. Ex Vitro Acclimatization

Before transplanting, the caps of the tissue culture bottles were first loosened and then a small amount of sterile water was added. The plants were acclimatized indoors for

24 to 48 h. The plantlets were gently removed using tweezers, and the residual culture medium was rinsed from the root necks with tap water. They were transplanted into four substrates: a mixture of vermiculite and peat soil (1:1); a mixture of peat soil, vermiculite, and perlite (1:1:1); a mixture of peat soil and vermiculite (1:2); and a mixture of peat soil, vermiculite, and perlite (1:2:1). Three groups of ten tissue-cultured plantlets were planted in each substrate for a total of thirty plants. They were watered thoroughly once and covered with clear plastic cups to maintain humidity for 3 days. The cups were then removed. The survival rate was counted after 30 days.

### 4.9. Statistical Analysis

The experimental data are expressed as the mean ± standard error (SE) and were analyzed using an ANOVA followed by Duncan's multiple range test in IBM SPSS Statistics 26.0. A significance level of $p < 0.05$ was established. Tables were created using Microsoft Excel 2010 and Microsoft Word 2020. The figures were produced in Adobe Photoshop 2020.

## 5. Conclusions

In this study, a complete rapid propagation technology for *Viburnum opulus* L. 'Roseum' was established, and the optimal culture conditions for different stages of tissue culture propagation were optimized. The best method for explant disinfection was found to be 75% ethanol for 30 s and 5% sodium hypochlorite for 5 min, resulting in a survival rate of 73.33%. The ideal medium for primary induction was WPM + 2.0 mg·L$^{-1}$ 6-BA + 0.15 mg·L$^{-1}$ IBA. The most effective proliferation scheme for adventitious buds was WPM + 1.0 mg·L$^{-1}$ 6-BA + 0.15 mg·L$^{-1}$ IBA + 25 g·L$^{-1}$ sucrose, with a proliferation efficiency of 7.17. The most suitable rooting medium was 1/2 WPM + 0.3 mg·L$^{-1}$ NAA + 0.3 mg·L$^{-1}$ AC + 25 g·L$^{-1}$ sucrose. The optimal transplant substrate was a 1:2:1 mixture of peat soil–vermiculite–perlite. The cultivation system effectively shortened the plantlet time and significantly improved the axillary bud induction rate and rooting rate of *Viburnum opulus* L. 'Roseum'.

**Author Contributions:** S.L. conceived and designed the experiments; Y.L. provided materials and funding; Y.N. and H.D. performed the experiments, analyzed the data, prepared the figures and tables, and contributed equally to this work; X.Z. and C.C. reviewed drafts of this paper. All authors have read and agreed to the published version of the manuscript.

**Funding:** This research received no external funding.

**Data Availability Statement:** All data are available in this manuscript.

**Conflicts of Interest:** In this article, the author, Yanhua Li, is based at Zhejiang Physical and Chemical Technology Co., Ltd. in Hangzhou, Zhejiang Province, and does not have a conflict of interest.

## References

1. Zhao, H.D.; Li, C.H.; Ai, X.R.; Xue, W.X.; Xu, L.X.; Xiang, Q.; Zhou, Y. Population Structure and Dynamic Characteristics of Five Wild *Viburnum* Plants in Southwest Hubei, China. *Guihaia* **2023**, *43*, 1747–1758.
2. Ma, J.; Li, F. A Preliminary Study of Asexual Propagation Ways of *Viburnum macrocephalum*. *J. Sichuan For. Sci. Technol.* **2019**, *40*, 71–73+81.
3. Du, X.; Shen, X.H. Effects of Cadmium Stress on Physiological and Biochemical Indices of *Viburnum odoratissimum* and *V. tinus* Seedlings. *Chin. J. Ecol.* **2010**, *29*, 899–904.
4. Bi, B.; Liu, Y.C.; Chen, Q.; Zhou, Z.; Zhang, X.X.; Sun, H. Heavy Metal Accumulation Capability of Ten Broadleaved Tree Species to As, Hg, Pb, Cd and Cr. *J. West China For. Sci.* **2012**, *41*, 79–83.
5. Chen, B.H.; Li, J.M.; Fan, H.H.; Zhang, J.; She, X.H.; Wang, S.F. Cutting Propagation Technique of *Viburnum macrocephalum* Fort. *Bangladesh J. Bot.* **2017**, *46*, 233–236.
6. Bi, X.Y. Study Breeding Technologies of *Viburnum sargenti* Koehne. Master's Thesis, Northeast Forestry University, Harbin, China, 2018.

7. Wijerathna-Yapa, A.; Hiti-Bandaralage, J. Tissue Culture—A Sustainable Approach to Explore Plant Stresses. *Life* **2023**, *13*, 780. [CrossRef] [PubMed]
8. VirVirginia, H.; Patricia, M.H. In Vitro Propagation of *Viburnum opulus* "Nanum". *J. Environ. Hort.* **1985**, *3*, 41–45.
9. Yaseen, S.A. Micorpropagation of *Viburnum opulus* (Roseum) by Using Single Nodes. *Iraqi J. Agric. Sci.* **2022**, *53*, 1388–1396. [CrossRef]
10. Bhadrawale, D.; Mishra, J.P.; Mishra, Y. An Improvised in Vitro Vegetative Propagation Technique for *Bambusa tulda*: Influence of Season, Sterilization and Hormones. *J. For. Res.* **2018**, *29*, 1069–1074. [CrossRef]
11. Wen, T.C.; Qian, Z.Y.; Li, B.Y. Research Progress on Explant Sterilization Technology of Plant Tissue Culture. *Mod. Agric. Sci. Technol.* **2024**, *13*, 72–76.
12. Tian, H.; Hou, N. Study on Tissue Culture and Rapid Propagation Technology System of *Polygonatum sibiricum* Red. *J. Nanjing Norm. Univ. Nat. Sci. Ed.* **2020**, *43*, 129–135.
13. Jian, D.; Gao, Y. Studies on the Technology of Primary Culture of *Viburnum sargenti* In Vitro. *J. Jilin Agric. Sci. Technol. Univ.* **2015**, *24*, 1–3.
14. Wang, H.B.; Ding, D.; Zheng, L.J.; Wang, J.Y.; Zheng, L.J.; Guo, Y.C. Optimization of Tissue Culture and Rapid Propagation Technology System of *Viburnum sargentii*. *Jiangsu Agric. Sci.* **2021**, *49*, 59–63.
15. Wu, J.; Fa, Y.L.; Zhu, C.Y.; Pang, X.W.; Wang, X.H. Microsomal Propagation of *Viburnum sargentii*. *J. Shandong For. Sci. Technol.* **2012**, *42*, 91–92.
16. Zhen, X.H.; Hu, H.Y.; Xia, Y.S. Study on Tissue Culture of *Viburnum opulus* L. *Anhui Agric. Sci. Bull.* **2010**, *16*, 57–59.
17. Guan, Q.Z.; Dong, X.C.; Dong, R.; Jiao, H.J.; Wang, H.W.; Wei, S.W. Study on Tissue Culture Rapid Propagation and Tissue Micro-Cuttage Technology of Pyrus Betulifolia. *North. Hortic.* **2024**, *20*, 23–29.
18. Wang, Z.H.; Li, Q.; Zhang, S.F.; Deng, X.M.; Wu, A.M. Establishment and Optimization of Tissue Culture and Regeneration System of *Ormosia henryi*. *Jiangsu Agric. Sci.* **2024**, *52*, 50–55.
19. Li, Z.J.; Cao, S.J.; Zhou, W. Tissue Culture and Rapid Propagation of *Vaccinium ashei* 'Tifblue'. *North. Hortic.* **2022**, *09*, 33–39.
20. Du, S.R.; Gu, T.T.; Pan, L.; Yang, W.X.; Jiang, C.Y. Studies on Rapid Propagation of *Viburunm farreri* in Tissue Culture. *J. Hebei For. Sci. Technol.* **2011**, *05*, 5–7.
21. Cai, N.; Wang, X.M.; Li, Y.X.; Zeng, H.J.; Qiao, Z.Q. Establishment of a Tissue Culture and Rapid Propagation System of *Lagerstroemia indica* 'Xiaoming 1'. *Chin. Agric. Sci. Bull.* **2016**, *32*, 22–27.
22. Luo, M.W. Study on Rapid Propagation and Polyploid Induction of *Platostoma palustr*. Master's Thesis, Guangxi Normal University, Guilin, China, 2023.
23. Bi, X.Y.; Li, S.J. Establishment of Tissue Culture and Regeneration System of *Viburnum sargentii*. *Hunan Agric. Sci.* **2018**, *4*, 1–4+12.
24. Liu, X.G.; Shan, M.H.; Su, S.; Wang, Y.Q.; Zhao, Y.; Dong, J.X. Optimization and Transplanting of Seeding Rooting Medium of *Armeniaca sibirica*. *Mol. Plant Breed.* **2020**, *18*, 1999–2005.
25. Pan, L.M.; Huang, Q.F.; Chen, Q.P.; Wan, L.Y.; Fu, J.E.; Wei, S.G. Tissue Culture of *Aristolochia fordiana* Hemsl. *Chin. J. Trop. Agric.* **2021**, *41*, 73–77.

*plants*

MDPI

*Article*

# Establishment of a Highly Efficient In Vitro Regeneration System for *Nandina domestica* 'Firepower'

**Xin Zhao [1,†], Hao Dong [1,†], Yanhua Li [2], Xinxin Zhang [1], Yajing Ning [1], Chengpeng Cui [1] and Shujuan Li [1,*]**

1 State Key Laboratory of Tree Genetics and Breeding, Northeast Forestry University, Harbin 150040, China; zhxin@nefu.edu.cn (X.Z.)
2 Zhejiang Physical and Chemical Technology Co., Ltd., Hangzhou 310000, China
* Correspondence: lishujuan@nefu.edu.cn
† These authors contributed equally to this work.

**Abstract**

*Nandina domestica* 'Firepower' is one of the most popular colorful foliage species in landscaping. However, it is currently propagated mainly by seeding and cuttings, with a low reproduction coefficient, hindering the cultivation of this species. Therefore, establishing an in vitro regeneration system would be beneficial for the industrialized production of *Nandina domestica* 'Firepower'. In this study, an ex vivo regeneration system was established using the direct organogenesis pathway. In early April, the new shoots of *Nandina domestica* 'Firepower' were selected, and the stem segments of 1~2 cm were cut as the disinfection materials for the explants. The optimal formulation for inducing axillary shoots was 1/2 MS + 1.5 mg $L^{-1}$ 6-benzylaminopurine (BA) + 0.3 mg $L^{-1}$ indole-3-butric acid (IBA). The optimal formulation for the differentiation and proliferation of axillary shoots was 1/2 MS + 1.5 mg $L^{-1}$ BA + 0.01 mg $L^{-1}$ IBA with a multiplicity of proliferation of 9.22. We determined that the rooting of axillary shoots required a combination of IBA, naphthalene acetic acid (NAA), and activated carbon (AC). The optimal formulation for rooting was 1/2 MS + 0.2 mg $L^{-1}$ NAA + 0.3 mg $L^{-1}$ IBA + 0.2 mg $L^{-1}$ AC. After a two-day hardening period for tissue-cultured plantlets, a substrate consisting of peat soil, vermiculite, and perlite at a ratio of 2:2:1 was determined to be the optimal cultivation formulation. This system provides a framework for the industrialized production of *Nandina domestica* 'Firepower'.

**Keywords:** *Nandina domestica* 'Firepower'; plant tissue culture; garden ornamental plants; rapid propagation

## 1. Introduction

*Nandina domestica* 'Firepower' is a cultivated variety of *Nandina domestica* Thunb in the Berberidaceae family and the genus *Nandina*. It is an evergreen shrub with dark green leaves that often display a red hue during the winter months. The species Nandina domestica is indigenous to North China, Central China, East China, and parts of South China. Additionally, the plant exhibits a number of beneficial characteristics, including drought resistance, wind resistance, high-temperature resistance, cold resistance, snow pressure resistance, and strong resistance to pests and diseases. Presently, *Nandina domestica* Thunb in China is primarily concentrated in Jiangsu, Zhejiang, Anhui, Jiangxi, Hunan, Hubei, Sichuan, Shaanxi, and Hebei, with a notable presence in various horticultural settings [1,2]. *Nandina domestica* 'Firepower' thrives in a warm, moist, well-ventilated, semi-shaded environment, preferring sandy loams rich in humus [3]. *Nandina domestica*

'Firepower' is a short plant with short nodes [4]. As temperatures decline in autumn, the foliage undergoes a transformation, shifting from its verdant hue to a vibrant red. This striking seasonal change can persist until the following spring, offering a compelling ornamental quality [5]. As society progresses and living standards improve, people have increasingly high expectations of the quality of the environment, particularly with regard to plants. The public is becoming increasingly interested in plants of different colors, leading to the development of some colorful varieties [6]. *Nandina domestica* 'Firepower' is used in garden landscapes such as flower beds and flower mirrors for its compact growth and the fact that its leaves turn red leaves in winter, providing vibrant color for winter scenery [7].

Nandina domestica 'Firepower' can be propagated through two methods: sexual reproduction (seeding) and asexual reproduction (ramet division and cutting). However, both methods exhibit low propagation efficiencies and extended growth cycles, making them susceptible to environmental influences. Specifically, the germination rate of seeding is only 60% to 70%, and it takes 2 to 3 years for seedlings to develop into mature plants suitable for landscaping or ornamental purposes. Ramet division is typically performed every 3 to 4 years; excessive division can negatively affect plant growth and fruiting. The rooting success rate of cuttings is approximately 70% to 80%, with strict requirements for environmental conditions such as humidity and temperature [8,9]. Plant tissue-culture technology offers many advantages, such as high reproduction efficiency, the stable growth of seedlings, and unrestricted reproduction. The use of plant tissue culture can enable the rapid propagation of the *Nandina domestica* 'Firepower' to meet market demand. To date, various cultivars of *Nantian bamboo* have been successfully propagated in vitro, including the "Lemon yellow" and "Habo" varieties [10]. In 2007, the first study on the micropropagation of *Nantian bamboo* was conducted, utilizing stem segments from that year's shoots as explants to identify the optimal medium for germination, redifferentiation, and rooting [11]. In 2010, Li Hui developed an in vitro culture system using sterilized vaccine stem segments as explants, investigating the impacts of different basal media, plant growth regulators, and activated carbon on plant regeneration [5]. However, in that in vitro culture system, the proliferation efficiency of the adventitious buds was low.

Therefore, with the aim of attaining higher reproductive efficiency, a stable and efficient in vitro culture system was developed on the basis of vigorous-growing and pest- and disease-free materials.

## 2. Results

### 2.1. Basal Medium Selection

As shown in Table 1 and Figure 1, the type of medium had a significant effect on the formation and growth of axillary shoots in *Nandina domestica* 'Firepower'. When inoculated to the 1/2 Murashige and Skoog (1/2 MS) medium, the axillary shoots' leaves were tender and green with robust growth (Figure 1a,d); when inoculated with Murashige and Skoog (MS) medium, the axillary shoots' leaves were tender and green, but the leaf surface was curled, the base was browned, and the medium was yellowed (Figure 1b,e); and when inoculated with woody plant medium (WPM), the axillary shoots leaves were yellowed, and growth was wilted (Figure 1c), (Figure 1f). Therefore, considering the induction of axillary shoots and plant growth, 1/2 MS was chosen as the optimal basal medium in this experiment.

**Table 1.** Effect of medium types on the growth of axillary shoots.

| Group | Type of Medium | Plant Growth |
|---|---|---|
| 1 | 1/2 MS | Tender green leaves and robust growth of axillary shoots in tissue culture |
| 2 | MS | Tissue-cultured axillary shoots with tender green leaves but with curled leaves and browning at the base. |
| 3 | WPM | Yellowish and shriveled leaves of tissue-cultured axillary shoots |

**Figure 1.** Effect of medium type on axillary shoots' initiation. (**a**) 1/2MS; (**b**) MS; (**c**) WPM. (**d**) The culture was maintained in a 1/2 MS medium for a period of 30 days. (**e**) The culture was maintained in a MS medium for a period of 30 days. (**f**) The culture was maintained in a WPM medium for a period of 30 days.

### 2.2. Induction of Axillary Shoots from Tender Stem

As shown in Table 2, the concentrations of 6-benzylaminopurine (BA) and indole-3-butyric acid solution (IBA) had significant effects on the induction of axillary shoots in *Nandina domestica* 'Firepower'. When the concentration of IBA was 0.5 mg-L$^{-1}$, the time required for inducing the axillary shoots was gradually shortened with the increase in the BA concentration. When the concentration of IBA was increased, the time required for inducing axillary shoots became longer at the same concentration of BA. Comparing the growth of axillary shoots and the time required to induce axillary shoots, we found that high concentrations of IBA inhibited the induction of axillary shoots in *Nandina domestica* 'Firepower'; therefore, after a comprehensive comparison, 1.5 mg-L$^{-1}$ BA and 0.3 mg-L$^{-1}$ IBA were found to be the optimal choices for axillary shoots' induction.

**Table 2.** Effect of PGR and concentration on the induction of axillary shoots.

| Groups | BA Concentration (mg L $^{-1}$) | IBA Concentration (mg L $^{-1}$) | Sprouting Time/Day | Growth State of Buds |
|---|---|---|---|---|
| 1 | 0.5 | 0.3 | 13.4 ± 0.90 a | Leaf blades young and green but slightly curled on the leaf surface |
| 2 | 1 | 0.3 | 9.84 ± 0.65 d | Tender green leaves but foliage curled |
| 3 | 1.5 | 0.3 | 7.08 ± 0.38 e | Tender green leaves and strong growth |
| 4 | 0.5 | 0.5 | 11.6 ± 0.24 c | Leaves slightly yellowed |
| 5 | 1 | 0.5 | 12.36 ± 0.88 b | Yellowing of leaves |
| 6 | 1.5 | 0.5 | 12.48 ± 0.50 b | Basal browning |

Means followed by the same letters in rows are not significantly different at $p \leq 0.05$.

### 2.3. Proliferation of Axillary Shoots

As shown in Table 3 and Figure 2, the concentrations of BA and IBA had significant effects on the proliferation of axillary shoots. When the concentration of IBA was constant, upon increasing the concentration of BA, the proliferation rate of axillary shoots of tended to increase; when the concentration of BA was constant, with the increase in the concentration of IBA, the proliferation rate of the axillary shoots tended to decrease. A comprehensive analysis of the proliferation rate and growth state of the axillary shoots revealed that the optimal medium for axillary shoots proliferation was 1/2 MS + 1.5 mg-L$^{-1}$ BA + 0.01 mg-L$^{-1}$ IBA, and the proliferation rate for axillary shoots was 9.22.

**Table 3.** Effects of different BA and IBA concentrations on the induction of bud clusters.

| Groups | BA Concentration (mg L $^{-1}$) | IBA Concentration (mg L $^{-1}$) | Multiplication Coefficient | Growth State of Buds |
|---|---|---|---|---|
| 1 | 1 | 0.01 | 3.98 ± 0.38 b | Average growth, curled leaves, few differentiated buds |
| 2 | 1 | 0.03 | 4.00 ± 0.67 b | Many differentiated buds, good growth, thick leaves |
| 3 | 1 | 0.05 | 3.00 ± 0.33 c | Average growth, curled leaves, few differentiated buds |
| 4 | 1.5 | 0.01 | 9.22 ± 0.84 a | Robust growth, normal leaf shape, maximum number of differentiated buds |
| 5 | 1.5 | 0.03 | 4.33 ± 0.57 b | Many differentiated buds, medium growth, browning at base |
| 6 | 1.5 | 0.05 | 2.66 ± 0.66 cd | Few differentiated buds, average growth, yellowing leaf margins |
| 7 | 1.5 | 0.1 | 1.88 ± 0.50 e | Few differentiated buds, with browning at the base |
| 8 | 1.5 | 0.3 | 1.44 ± 0.19 e | Differentiated buds, few in number, brownish at the base, partly vitrified |

Means followed by the same letters in rows are not significantly different at $p \leq 0.05$.

**Figure 2.** Effect of BA and IBA concentrations on shoot proliferation: (**a**) 1 mg-L$^{-1}$ BA + 0.01 mg-L$^{-1}$ IBA; (**b**) 1 mg-L$^{-1}$ BA 0.03 mg-L$^{-1}$ IBA; (**c**) 1 mg-L$^{-1}$ BA 0.05 mg-L$^{-1}$ IBA; (**d**) 1.5 mg-L$^{-1}$ BA + 0.01 mg-L$^{-1}$ IBA; (**e**) 1.5 mg-L$^{-1}$ BA + 0.03 mg-L$^{-1}$ IBA; (**f**) 1.5 mg-L$^{-1}$ BA + 0.05 mg-L$^{-1}$ IBA; (**g**) 1.5 mg-L$^{-1}$ BA + 0.1 mg-L$^{-1}$ IBA; (**h**) 1.5 mg-L$^{-1}$ BA + 0.3 mg-L$^{-1}$ IBA. (**i**) Subsequent cultures were cultivated on two occasions at a concentration of 1.5 mg L$^{-1}$ BA and 0.01 mg L$^{-1}$ IBA hormones.

### 2.4. Induction of Adventitious Roots

The impact of varying concentrations of naphthalene acetic acid (NAA), indole-3-butyric acid (IBA), and activated carbon (AC) on adventitious root growth was evaluated using the metrics presented in Table 4 and the data illustrated in Figure 3. The adventitious roots of tissue-cultured microshoots were primarily situated within the medium, with secondary roots becoming visible after approximately 20 days. Preliminary experiments indicated that the individual application of NAA or IBA to the medium led to a decrease in the number of roots formed, suggesting that these two plant growth regulators must act synergistically. When the concentration of NAA was held constant, an increase in the IBA concentration inhibited the development of adventitious roots. Conversely, at a fixed IBA concentration, the addition of low concentrations of NAA promoted the formation of adventitious roots. Taking into account the rooting rate, the number of roots, and their growth condition, the optimal rooting formula was determined to be 1/2 MS supplemented with 0.2 mg·L$^{-1}$ NAA, 0.3 mg·L$^{-1}$ IBA, and 0.2 mg·L$^{-1}$ AC, achieving a rooting rate of 88.67% and an average of 6.65 roots per plant.

**Table 4.** Effect of different concentrations of NAA, IBA, and AC on the rooting of axillary shoots.

| Groups | NAA Concentration (mg L$^{-1}$) | IBA Concentration (mg L$^{-1}$) | AC Concentration (mg L$^{-1}$) | Adventitious Roots Number | Rooting Rate | Grouped Microshoots and Root Growth Status |
|---|---|---|---|---|---|---|
| 1 | 0.2 | 0.2 | 0 | 0.00 ± 0.00 d | 0.00% | Microshoots are slow growing, with healing wounds at the base and no root formation. |
| 2 | 0.2 | 0.3 | 0.2 | 6.65 ± 0.59 a | 88.67% | Microshoots are strong, with a large number of roots and secondary root formation. |
| 3 | 0.2 | 0.4 | 0.4 | 1.66 ± 0.57 bcd | 27.67% | Microshoots are poor, with yellowing leaves and weak roots. |
| 4 | 0.2 | 0.5 | 0.5 | 1.66 ± 1.52 bcd | 11.33% | Microshoots are poorly grown, with deformed leaf blades and a low number of roots. |
| 5 | 0.3 | 0.2 | 0.2 | 5.36 ± 0.37 a | 55.67% | Microshoots are robust, with low root formation. |
| 6 | 0.3 | 0.3 | 0 | 0.00 ± 0.00 d | 0.00% | Microshoots are slow growing, with healing wounds at the base and no root production. |
| 7 | 0.3 | 0.4 | 0.5 | 1.83 ± 1.04 bc | 22.33% | Microshoots are strong, with low root production. |
| 8 | 0.3 | 0.5 | 0.4 | 1.33 ± 1.15 bcd | 11.33% | Microshoots are poorly developed, with deformed leaves and weak roots. |
| 9 | 0.4 | 0.2 | 0.4 | 2.66 ± 0.76 b | 33.00% | Microshoots are slow growing, with short roots. |
| 10 | 0.4 | 0.3 | 0.5 | 1.83 ± 0.76 bc | 22.33% | Microshoots are slow to grow, and fewer roots are produced. |
| 11 | 0.4 | 0.4 | 0 | 1.33 ± 2.30 bcd | 5.67% | Microshoots are poor, with healing wounds forming at the base. |
| 12 | 0.4 | 0.5 | 0.2 | 0.33 ± 0.57 cd | 5.67% | Microshoots poorly grown, and few roots are formed. |
| 13 | 0.5 | 0.2 | 0.5 | 1.00 ± 1.00 bcd | 16.67% | Microshoots are slow to grow, with a small number of roots. |
| 14 | 0.5 | 0.3 | 0.4 | 1.33 ± 1.15 bcd | 11.33% | Microshoots are slow growing, with weak roots. |
| 15 | 0.5 | 0.4 | 0.2 | 1.00 ± 1.73 bcd | 5.67% | Microshoots are poor, with yellowing leaves and a few roots forming. |
| 16 | 0.5 | 0.5 | 0 | 0.33 ± 0.57 cd | 5.67% | Microshoots are poorly grown, with yellowing leaves and healing at the base. |

Means followed by the same letters in rows are not significantly different at $p \leq 0.05$.

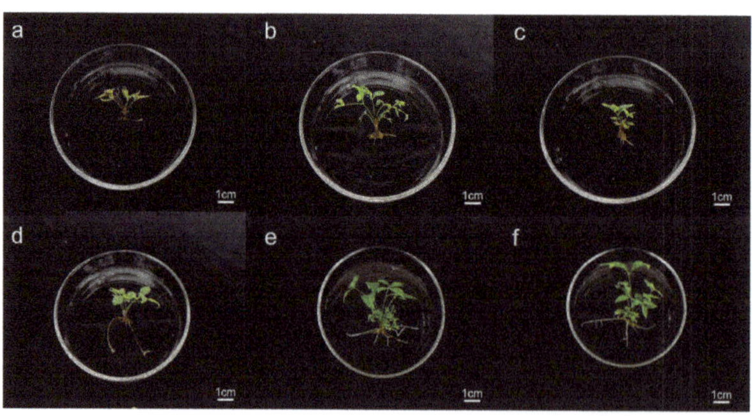


**Figure 3.** Effect of IBA, NAA, and AC concentrations on adventitious roots: (**a**) 0.2 mg-L$^{-1}$ NAA + 0.4 mg-L$^{-1}$ IBA + 0.4 mg-L$^{-1}$ AC; (**b**) 0.4 mg-L$^{-1}$ NAA + 0.2 mg-L$^{-1}$ IBA + 0.4 mg-L$^{-1}$ AC; (**c**) 0.4 mg-L$^{-1}$ NAA + 0.2 mg-L$^{-1}$ IBA + 0.4 mg-L$^{-1}$ AC; (**d**) 0.4 mg-L$^{-1}$ NAA + 0.3mg-L$^{-1}$ IBA + 0.5 mg-L$^{-1}$ AC; (**e**) 0.3 mg-L$^{-1}$ NAA + 0.2 mg-L$^{-1}$ IBA + 0.2 mg-L$^{-1}$ AC; (**f**) 0.2 mg-L$^{-1}$ NAA + 0.3 mg-L$^{-1}$ IBA + 0.2 mg-L$^{-1}$ AC.

### 2.5. Acclimatization

In this experiment, the impact of various substrates on the acclimatization of *Nandina domestica* 'Firepower' tissue-culture plantlets was investigated. Following a two-day acclimation period, the tissue-culture plantlets were transplanted into substrates composed of differing ratios of vermiculite, peat, and perlite. As shown in Table 5 and Figure 4, the plantlets initially cultivated in peat grew vigorously. However, as the cultivation time extended, there was a tendency for the edges of the leaves to dry out. After 30 days, the survival rate was only 3%. The plantlets cultivated in a 1:1 mixture of peat and vermiculite showed weak and slow growth after transplantation, and their survival rate was 26% after 30 days. In contrast, the plantlets grown in a 2:2:1 mixture of peat, vermiculite and perlite were vigorous and well-developed, with a survival rate of 86% after 30 days.

**Table 5.** Effects of different substrate proportions on plantlets.

| Group | Substrate Ratio | Survival Rate |
|---|---|---|
| 1 | peat soil | 0.03 ± 0.05 c |
| 2 | peat soil: vermiculite = 1:1 | 0.26 ± 0.05 b |
| 3 | peat soil: vermiculite: perlite = 2:2:1 | 0.86 ± 0.05 a |

Means followed by the same letters in rows are not significantly different at $p \leq 0.05$.

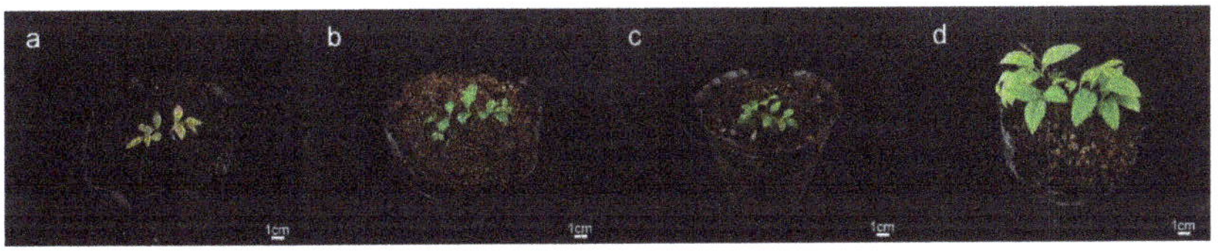


**Figure 4.** Growth of *Nandina domestica* 'Firepower' in different substrates: (**a**) peat soil; (**b**) peat soil: vermiculite = 1:1; (**c**) peat soil: vermiculite: perlite = 2:2:1; and (**d**) after 60 days incubation in a mixed substrate of peat soil: vermiculite: perlite of 2:2:1.

## 3. Discussion

*Nandina domestica* 'Firepower' is a shrub grown in gardens. It is typically planted in clusters or as a standalone specimen in flower beds and courtyards. Additionally, it can be utilized as ground cover and as a colorful accent plant. It is a highly sought-after material for flower beds and courtyards. Market demand for this plant is considerable [12]. The propagation of *Nandina domestica* 'Firepower' is typically achieved through dividing and cutting. However, this method is not without limitations. These include a relatively low propagation coefficient, the potential for the species characteristics to be compromised, and a lengthy cultivation period [13]. In recent years, the increasing demand for *Nandina domestica* 'Firepower' has rendered the traditional propagation method inadequate to meet market demand [11]. Nevertheless, the in vitro propagation of plant species could yield a considerable number of robust plantlets with stable growth patterns in a relatively brief period of time. In this study, we established a comprehensive system of plant tissue culture, encompassing axillary shoots' induction, shoot cluster induction, rooting, and transplantation. The choice of medium and the concentration of plant growth regulators have important effects on the induction and proliferation of axillary shoots in *Nandina domestica* 'Firepower'. According to Liu Zhong bing et al., the optimal combination for inducing axillary shoots of *Nandina domestica* 'Firepower' is MS + 1.0 mg-$L^{-1}$ 6 BA + 0.1 mg-$L^{-1}$ NAA + 20 g-$L^{-1}$ sucrose, which gave the highest induction rate of 74.4% [14]. Deng Yuying et al. observed that vitrification occurred in histo-cultured axillary shoots when the BA concentration reached 0.8 mg-$L^{-1}$. It is, therefore, thought that the concentration of BA in the culture medium should not be excessive [15]. However, our results show that the optimal medium for inducing axillary shoots of *Nandina domestica* 'Firepower' was 1/2 MS + 1.5 mg/L

BA + 0.3 mg/L IBA. The induction rate could reach 100%, and the generation time of axillary shoots was relatively short. Axillary shoots could be observed after 7 days of culture. Furthermore, high concentrations of BA were more conducive to the formation of axillary shoots. The discrepancy in the observed outcomes may be attributed to the fact that different varieties of nandina may necessitate varying concentrations of BA for optimal growth. Zhou Zhi jiang et al. demonstrated that the optimal proliferation medium for *Nandina domestica* 'Firepower' was MS + 1.0 mg-$L^{-1}$ BA + 0.1 mg-$L^{-1}$ NAA + 30 g-$L^{-1}$ sucrose, with a proliferation coefficient of 3.5 observed [16]. In a study of rapid propagation technology of *Nandina domestica* 'Firepower', Du Yong qin et al. demonstrated that MS + 1.0 mg of $L^{-1}$ BA + 0.1 mg-$L^{-1}$ IBA was the optimal proliferation medium, exhibiting the highest value-added coefficient of 4.12. In a separate study, Song Shuai jie et al. used young stem tips as the research object. They found that the optimal proliferation medium was MS + 1.0 mg-$L^{-1}$ BA + 0.1 mg-$L^{-1}$ NAA + 0.1 mg-$L^{-1}$ IAA + 1.0 g-$L^{-1}$ AC, which yielded the highest proliferation coefficient of up to 5.29 [17]. Our results show that the optimal medium for the proliferation of axillary shoots of *Nandina domestica* 'Firepower' was 1/2 MS + 1.5 mg/L BA + 0.01 mg/L IBA, with a proliferation coefficient as high as 9.22, which was significantly higher than the results of other studies on the proliferation of axillary shoots of *Nandina domestica* 'Firepower'. The observed variability in results may be attributed to the differing hormone types and ratios, which exert some influence on the growth of late-stage histo-cultured microshoots. In a rooting test conducted by Wang Chun et al. on bamboos (Nandina), the addition of IBA at a concentration of 0.25 mg L-1 resulted in a rooting rate of up to 87.33% [18]. Cai Peng et al. demonstrated that the rooting rate was 76.67% in a liquid medium comprising 1/2 MS + 0.2 mg $L^{-1}$ NAA + 0.2 mg $L^{-1}$ IBA + 4 g $L^{-1}$ AC [19]. We explored the influences of NAA, IBA, and AC on the rooting of micro-branches and the strengthening of microshoots. The results indicated that the combination of 1/2 MS + 0.2 mg/L NAA + 0.3 mg/L IBA + 0.2 g/L AC significantly enhanced the rooting rate of *Nandina domestica* 'Firepower', reaching up to 88%. The difficulties associated with rooting and the low survival rate of transplants represent significant challenges in the field of woody tissue-culture breeding. These obstacles impede the implementation and advancement of tissue-culture breeding technology in industrial production settings [20,21]. The addition of AC to the culture medium adsorbs harmful metabolites and promotes the generation of adventitious roots [22]. With regard to the transplantation of plantlets, Song Shuai jie and colleagues have demonstrated that the survival rate of transplanted plants can reach 94% when the transplantation substrate is a mixture of peat soil, vermiculite, and perlite at an 8:1:1 ratio [17] Zhang Jun lin et al. concluded that a 3:2 mixture of peat soil and perlite was more conducive to the growth and development of B. nantianensis [10]. We compared the influences of several substrate mixtures on the growth of *Nandina domestica* 'Firepower'. Ultimately, a 2:2:1 mixture of peat soil, vermiculite, and perlite was chosen as the cultivation substrate. Its survival rate can reach 86%. Subsequent studies will further explore the appropriate cultivation conditions for *Nandina domestica* 'Firepower'.

## 4. Materials and Methods

### 4.1. Plant Materials

*Nandina domestica* 'Firepower' was purchased from the Zhejiang flower and seedling market. Seven three-year-old *Nandina domestica* 'Firepower' plants were selected as raw materials. In early April, the new shoots of *Nandina domestica* 'Firepower' were selected, and the stem segments of 1~2 cm were cut as the main materials for the surface sterilization of explants.

### 4.2. Components of the Medium

The chemicals used were as follows: MS Base Salts with vitamins (PM519); 6-benzylaminopurine (BA) (1 mg·mL$^{-1}$), indole-3-butric acid solution (IBA) (1 mg·mL$^{-1}$), and naphthalene acetic acid (NAA) (1 mg·mL$^{-1}$) were from Phyto Technology Laboratories (Lenexa, KS, USA); sucrose (A502792), AGAR (A505255), and active charcoal (A502167) were from Sengon Biotech (Shanghai, China).

### 4.3. Culture Media and Conditions

In this research, three types of media, namely WPM, MS, and 1/2 MS, were utilized. The sucrose concentration of all these three media was 30 g/L, the agar concentration was 5.4 g/L, and the pH value was 5.8 ± 0.02. The types and concentrations of the plant growth regulators in the media employed at different cultivation stages were diverse. All media were sterilized at 121°C for 20 minutes prior to their utilization. The culture temperature was 25 ± 1 °C, the light intensity was 40–50 μmol·m$^{-2}$·s$^{-1}$, and a light/dark cycle of 16/8 h was used.

### 4.4. Sterilization of Explant

In early April, the new shoots of *Nandina domestica* 'Firepower' were harvested, and 1–2 cm long stem segments were excised to serve as explants. They were initially rinsed under running water for 120 min (given that the internodes of *Nandina domestica* 'Firepower' are extremely short, a prolonged rinsing period is necessary to eliminate certain external contaminant sources) Subsequently, they were transferred into tissue-culture flasks. All subsequent procedures were conducted in a vertical laminar flow cabinet. The explants were initially surface-sterilized with 75% ethanol for 30 s, followed by rinsing 3–5 times with sterile water to eliminate the residual ethanol. Subsequently, disinfection was continued using 2% sodium hypochlorite for 10 min, after which the explants were rinsed again 3–5 times with sterile water. Eventually, the stem segments were dried on sterile filter paper to prepare for the induction of axillary shoots.

### 4.5. Basal Culture Medium Selection

The stem segments obtained from the sterilized materials in the previous stage were used as the experimental materials. These stem segments were inoculated into three different culture media: WPM, MS, and 1/2MS, and cultured for 30 days to monitor their growth. Each bottle contained three stem segments, with a total of 10 bottles per treatment, and each treatment was replicated three times. After the 30-day culture period, the growth status of the stem segments was evaluated and recorded.

### 4.6. Induction of Axillary Shoots from Tender Stem

Sterilized stem segments were used as experimental materials and transferred to media containing BA (0.5, 1.0, or 1.5 mg L$^{-1}$) and IBA (0.3 or 0.5 mg L$^{-1}$) to induce the formation of axillary shoots. Three to four stem segments were inoculated into each bottle. For each combination, a total of 30 bottles were inoculated (10 bottles per replicate, with three replicates). After a 30-day culture period, the time of emergence and growth status of the axillary shoots were recorded. Detailed information on these combinations is provided in Table 2.

### 4.7. Subculture of Axillary Shoots

The axillary shoots induced in the previous stage were transferred to media containing BA (1.0 or 1.5 mg L$^{-1}$) and IBA (0.01, 0.03, 0.05, 0.1, or 0.3 mg L$^{-1}$) for subculture. Each bottle contained three to four stem segments, with a total of 10 bottles inoculated per treatment, and each treatment was replicated three times. After a 30-day incubation period, the proliferation and growth status of the shoots were carefully observed and recorded. The proliferation coefficient was calculated by dividing the total number of new shoots formed by the number of initially inoculated buds.

### 4.8. Rooting

Microshoots measuring between 1.5 and 2.5 cm in height were meticulously selected, separated from the shoot clusters, and subsequently cultured in a 1/2 MS medium supplemented with varying concentrations of IBA, NAA, and AC. Each bottle contained three stem segments, with a total of 10 bottles inoculated per treatment, and each treatment was replicated three times. After a 30-day culture period, the number, length, and time required for root formation were carefully documented. The rooting efficiency was quantified as the ratio of microshoots that successfully developed roots to the total number of microshoots inoculated into the medium.

### 4.9. Ex Vitro Acclimatization

Plantlets with well-developed roots were selected from the rooting culture for acclimatization. An appropriate amount of distilled water was added to the tissue-culture flasks, which were left open gradually over 1 to 2 days to allow the plantlets to acclimate to the external environment during this period. After acclimatization, the plantlets were carefully removed from the flasks, and their roots were gently washed to remove any residual culture medium. The plantlets were then planted in substrates containing different proportions of vermiculite, peat soil, and perlite. Thirty plantlets were planted in each substrate type (10 plantlets per replicate, with three replicates). After a 30-day growth period, the survival rate and growth conditions of the plantlets were observed and recorded.

### 4.10. Statistical Analysis

The experimental data are presented as the mean $\pm$ standard error (SE) and were analyzed through SPSS 25.0 using ANOVA followed by Duncan's multiple range test. The significance level was set at $p < 0.05$. Tables were constructed via Microsoft Excel 2010 and Microsoft Word 2021, and the figures were generated using PS 2020.

## 5. Conclusions

In this study, we established an in vitro propagation system for *Nandina domestica* 'Firepower' via direct organogenesis. In early April, the new shoots of *Nandina domestica* 'Firepower' were selected, and the stem segments of 1~2cm were cut as the main materials for the surface sterilization of explants. The optimal protocol for explant sterilization was found to be a two-hour rinse with running water, a 30 s treatment with 75% ethanol, and a 10 min treatment with 2% NaOCl. The optimal medium was identified as 1/2 MS basal medium, and the most effective formulation for the induction of axillary shoots was 1/2 MS medium supplemented with 1.5 mg $L^{-1}$ BA and 0.3 mg $L^{-1}$ IBA. The shoots of tissue-culture seedlings were used to induce axillary shoots, and the optimal concentrations for subculture and proliferation were found to be 1.5 mg $L^{-1}$ BA and 0.01 mg $L^{-1}$ IBA. Rooting was performed using 1/2 MS + 0.2 mg $L^{-1}$ NAA + 0.3 mg $L^{-1}$ IBA + 0.2 mg $L^{-1}$ AC. Finally, the optimal substrate was found to be vermiculite, peat soil and perlite in a 2:2:1 ratio. Nevertheless, it is widely acknowledged that the genotype significantly influences plant regeneration, and certain variations exist among different genotypes. Hence, in future studies, we will conduct research on other genotypes of this plant to enrich the in vitro culture system of this plant.

**Author Contributions:** S.L. conceived and designed the experiments; Y.L. provided materials and funding; X.Z. (Xin Zhao) and H.D. performed the experiments, analyzed the data, prepared the figures and tables, and they contributed equally to this work; X.Z. (Xinxin Zhang), Y.N. and C.C.

participated in the experimental validation. All authors have read and agreed to the published version of the manuscript.

**Funding:** This research received no external funding.

**Data Availability Statement:** All data are available in the manuscript.

**Conflicts of Interest:** Author Yanhua Li was employed by the company Zhejiang Physical and Chemical Technology Co., Ltd. The remaining authors declare that the research was conducted in the absence of any commercial or financial relationships that could be construed as a potential conflict of interest.

## References

1. Tang, L.; Liu, Y.Q. Germ Plasm Resource and Application of *Nandina domestica*. *J. Hunan Ecol. Sci.* **2007**, *3*, 4–6.
2. Long, J.F. Scientific development and utilization of Hunan garden plant resources. *For. Ecol.* **2006**, *4*, 15.
3. Jiang, Z.P.; Wang, F.Y.; Xu, Z.D.; Zeng, K.; Shi, Y.G. Preliminary study on in vitro preservation technique of *Nandina domestica* 'Fire power'. *J. Jiangsu For. Sci. Technol.* **2010**, *37*, 6–8.
4. Song, G.; Zhu, Y. Tissue Culture and Large-Scale Production of *Nandina domestica* 'Firepower'. *Plant Physiol. J.* **2010**, *46*, 157–158.
5. Li, H. In vitro fast propagation of the colorful leaf tree species *Nandina domestica* 'Firepower'. *Jiangsu Agric. Sci.* **2010**, *1*, 74–76.
6. Yu, J.; Shen, X.T. Development and Application of *Nandina domestica* 'Firepower'. *Chin. Hortic. Abstr.* **2009**, *25*, 153.
7. Jiao, L.L.; Li, X.P.; Liu, X.C. Characteristics and garden application of the foliage plant *Nandina domestica* 'Firepower' Nandina. *Mod. Rural. Sci. Technol.* **2021**, *6*, 69.
8. Liu, G.H.; Chen, Y.; He, Y.P.; Yang, S.H. Sowing and Propagation Techniques of *Nandina domestica*. *Hortic. Seed* **2018**, *38*, 18–19.
9. Yang, Y.H. Breeding Techniques of *Nandina domestica*. *China Flowers Hortic.* **2017**, *4*, 33–35.
10. Xie, Y.Q.; Ning, L.; Li, X.H.; Chen, H.F.; Liao, L.Y.; Zhao, G.D. Study on tissue culture rapid propagation technology of *Calliandra haematocephala*. *Trop. For.* **2019**, *47*, 25–28.
11. Zeng, Y.P.; Shao, C.R.; Wang, C.; Xu, S.Y. Fast propagation and production of red-leafed Nandina. *China Flowers Hortic.* **2013**, *6*, 36–38.
12. Zeng, H.J.; Zhang, M.L.; Zou, C.J. Study on the tissue culture propagation technique of *Nandina domestica* 'Firepower'. *Mod. Agric.* **2015**, *9*, 8–10.
13. Du, Y.Q.; Ni, L.J.; Wang, Y.Q.; Zhang, Y.C.; Liu, X.W. Studies on introduction and propagation technology of a new ornamental forest variety *Nandina domestica* 'Fire power'. *Acta Agric. Shanghai* **2007**, 3, 38–41.
14. Liu, Z.B.; Meng, Y. Optimization of Primary Culture Conditions in vitro of *Nandina domestica* Thunb. *North. Hortic.* **2015**, *5*, 93–96.
15. Deng, Y.Y.; Lv, X.L. Study on Tissue Culuture of *Nandina domestica* 'Firepower'. *North. Hortic.* **2010**, *20*, 141–143.
16. Zou, Z.J.; Chen, M.L.; Yang, P.J. Establishment and Optimization of Tissue Culture System for *Nandina domestica* 'Fire power'. *J. Agric. Sci. Technol.* **2019**, *21*, 164–168.
17. Song, S.J.; Huang, K.; Qian, R.N.; Liu, F.; Lin, L.; Chen, J.K. Study on tissue culture and rapid propagation system of *Nandina domestica* Firepower. *J. Zhejiang Agric. Sci.* **2020**, *61*, 2062–2065.
18. Wang, C.; Zhang, J.L.; Fang, W.F.; Wang, Y.; Fang, H. Techniques of fast-growth reproduction of *Bambusa alba*. *J. For. Eng.* **2011**, *25*, 85–88.
19. Cai, P.; Su, H.T.; Li, L.Y.; Liu, H.Y.; Yao, H.W. Study of in Vitro Culture of *Nandina domestica* Thunb. *J. Hebei Norm. Univ. (Nat. Sci.)* **2016**, *40*, 66–72.
20. Zhang, L.L.; Zhang, M.M. Application and limitations of woody plant tissue culture technology in forestry research and production. *Beijing Agric.* **2015**, *9*, 74–75.
21. Wang, H.Y.; Li, L.F.; Chen, Y.; Chen, N.M. Study on Rapid Propagation in vitro of Two Cultivars of *Nandina domestica*. *Subtrop. Plant Sci.* **2014**, *43*, 255–258.
22. Zhang, J.L.; Fan, Q.; Hu, Y.C.; Yang, H.; Hu, S.Q.; Shen, B.C. *Nandina domestica* 'Harbour Dwarf': Tissue Culture and Rapid Propagation. *J. Agric.* **2021**, *11*, 63–68+89.

*plants*

MDPI

*Article*

# Sex-Specific Responses of Sexual Reproduction, Clonal Reproduction, and Vegetative Growth to Environmental (Biotic and Abiotic) Factors in the Clonal Dioecious Plant *Acer barbinerve*

**Dan Liao [1], Jingjing Lei [1], Yingni Wang [1], Yuxin Bao [1], Xinna Zhang [2] and Juan Wang [1,*]**

[1] Key Laboratory for Forest Resources & Ecosystem Processes of Beijing, School of Ecology and Nature Conservation, Beijing Forestry University, Beijing 100083, China; liaodan@bjfu.edu.cn (D.L.); 18294324528@163.com (J.L.); wangyingni08042@163.com (Y.W.); baoyx0283@163.com (Y.B.)

[2] College of Forestry, Beijing Forestry University, Beijing 100083, China; zhangxinna@bjfu.edu.cn

* Correspondence: claire_wangj@163.com; Tel.: +86-138-1027-7963

**Abstract**

Sexual dimorphism in dioecious plants serves as a critical adaptive strategy in complex environments. This study systematically investigated the effects of topographic factors (elevation, slope, aspect, and convexity), soil nutrients (C, N, P), and interspecific competition intensity on the reproductive strategies and vegetative growth of the clonal dioecious plant, *Acer barbinerve*. Using Spearman's correlation analysis, multiple regression models, and PLS-PM path models, key findings include the following: (1) female sexual reproduction biomass showed a significant positive correlation with the topography principal component (topo_PC1), with a notable gender–topography interaction, whereas male sexual reproduction was negatively regulated by elevation; (2) clonal reproduction in both sexes was significantly suppressed by interspecific competition, but females additionally exhibited positive topographic responses; and (3) male vegetative growth was significantly impacted by environmental stress, while females maintained relative stability. These results demonstrate that females optimize reproductive investment through topography-mediated resource acquisition, whereas males are more susceptible to resource competition constraints. This sex-specific adaptive strategy corroborates the dimorphic niche hypothesis, highlighting how environmental heterogeneity drives divergent life history allocations in dioecious species. The findings provide novel insights into the ecological mechanisms underlying sexual dimorphism and inform gender ratio management in ecological restoration practices.

**Keywords:** environmental adaptation; sexual dimorphism; clonal plant; dioecy

## 1. Introduction

Dioecious plants play an important role in terrestrial ecosystems [1]. Their sexual dimorphism reflects adaptive traits associated with reproductive roles [2,3]. The ecological, morphological, and physiological differences between sexes in dioecious plants are often attributed to varying reproductive costs of each sex and are closely related to the trade-offs assigned to reproductive and other plant functions [4,5]. Throughout their life cycle, plants are continuously influenced by both biotic and abiotic factors [6], and different sexes often respond differently to environmental changes. For example, females of *Populus euphratica*


https://doi.org/10.3390/plants14040596

are more sensitive to drought and salinity than males [7], while male *Salix matsudana* exhibit stronger salt tolerance [8]. Therefore, understanding how environmental factors (both biotic and abiotic) affect the plant growth and reproduction of dioecious plants is crucial [9–11]. However, studies on the impact of environmental factors on dioecious plants still have two main limitations: a lack of systematic comparisons and limited consideration of multifactorial interactions [12,13].

According to the dimorphic niche hypothesis, differences in reproductive costs and allocation trade-offs between females and males may lead to their different responses to environmental gradients [14], and abiotic factors related to geography also play a role in modifying sexual dimorphism in plant life history [15]. Topographic features such as slope, aspect, and elevation are considered the primary factors causing spatial variations [16], and they promote the formation of environmental heterogeneity by regulating conditions such as soil moisture, light, and microclimate [17–19]. This heterogeneity further leads to differences in vegetative growth and reproduction processes [16,20,21]. In addition, plant growth and reproduction are limited by the availability of key elements (e.g., carbon, nitrogen, and phosphorus) in the environment [22,23]. For dioecious plants, there are differences in resource requirements between males and females [24]. For example, male plants may require more nitrogen for pollen production during flowering, whereas female plants require more carbon for seed production during fruiting [25]. Thus, sexual dimorphism exists in dioecy in response to topography and soil nutrients during growth and reproduction.

De Lisle demonstrated that resource competition drives the evolution of sexual dimorphism [26], which implies that there is a gender difference in the response of the two sexes to interspecific competition [27,28]. Plants compete for resources needed for growth and reproduction through interspecific competition in environments with limited resources. This competition usually drives plants to make morphological and physiological adaptations [29,30]. Different plant species respond differently to resource competition, and clonal plants in particular are more complex regarding interspecific competition due to their dual ability for both clonal growth and sexual reproduction. The resource allocation of clonal dioecious plants is often strongly influenced by interspecific competition pressures [31,32]. Therefore, studying the adaptation of dioecious plants under different interspecific competition intensities, especially the manifestation of sexual reproduction, clonal reproduction, and vegetative growth, is important for understanding plant growth and reproductive strategies in complex environments.

This study focuses on *Acer barbinerve* and systematically investigates the combined effects of abiotic factors (such as topography and soil nutrients) and biotic factors (such as interspecific competition) on its sexual reproduction, clonal reproduction, and vegetative growth. *Acer barbinerve* is a typical clonal dioecious plant in the Aceraceae family and is one of the foundational species in the temperate forests of China. It has high species diversity in its local area and is also one of the dominant understory species [33,34]. The sex of this plant can be easily distinguished during the flowering period (April to May), with the fruiting period occurring in July to August. As a root-suckering clonal plant, *Acer barbinerve* expands horizontally through underground rhizomes and forms new buds at certain nodes, which eventually develop into independent individuals. Due to the small spacing between its ramets, they can grow closely together, making it easier to identify the genets in the field. This study will address the following scientific questions: (1) Do abiotic factors have sexually dimorphic effects on *Acer barbinerve*? (2) Do biotic factors have sexually dimorphic effects on *Acer barbinerve*? (3) Are the differences in sexual dimorphism consistent across sexual reproduction, clonal reproduction, and vegetative growth?

## 2. Results

### 2.1. The Effects of Biotic and Abiotic Factors on Sexual Reproduction

In the Spearman's correlation analysis, slope and sinA are significantly positively correlated with sexual reproduction in females, while sexual reproduction in males is significantly negatively correlated with elevation and interspecific competition intensity (Figure 1). Other environmental factors, although showing some correlation, do not exhibit significant relationships. The multiple regression model for sexual reproduction (Figure 2) indicates that gender ($\beta = -1.30$, $p < 0.001$), topo_PC1 ($\beta = 0.28$, $p < 0.01$), and soil_PC1 ($\beta = 0.26$, $p < 0.05$) have significant effects on sexual reproduction biomass. Additionally, a significant interaction effect between gender and topo_PC1 was observed ($\beta = -0.29$, $p < 0.05$), suggesting that the influence of topo_PC1 on sexual biomass is weaker in males compared to females. A marginal interaction effect between gender and soil_PC1 was also found ($p < 0.1$). In the PLS-PM models (Figure 3), the environmental factor exhibits a significant negative effect on female sexual reproduction ($\beta = -0.60$, $p < 0.01$), while its negative impact on male sexual reproduction is not statistically significant ($\beta = -0.33$, $p > 0.05$).

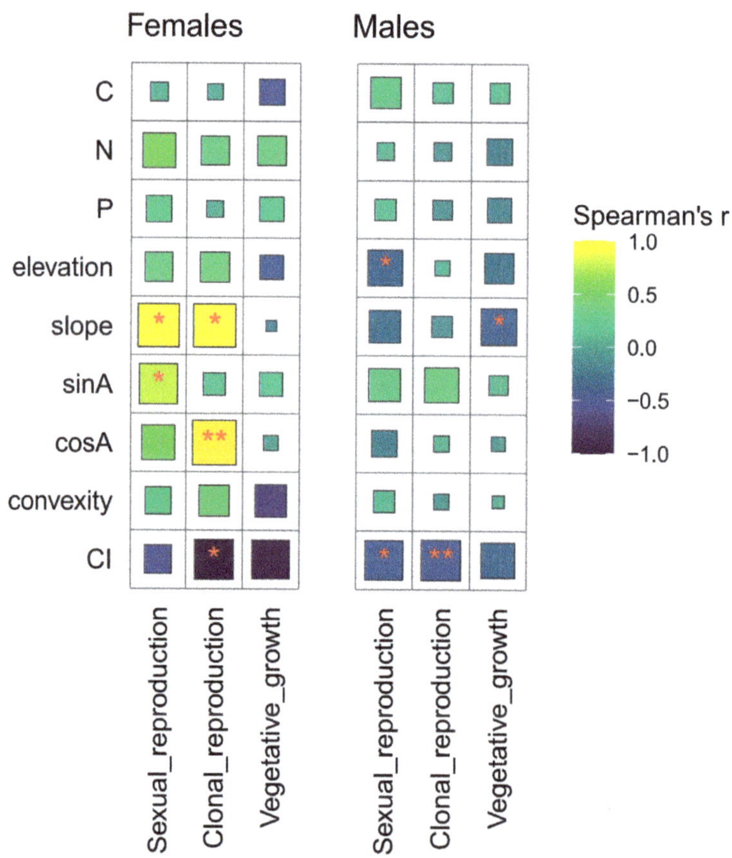


**Figure 1.** Correlations between sexual reproduction, clonal reproduction, vegetative growth, and their explanatory variables. The size of the squares indicates the significance of Spearman's correlation between explanatory variables, and the colors represent the different strengths of the correlations. C, soil carbon content; N, soil nitrogen content; P, soil phosphorus content; sinA, east aspect; cosA, north aspect; CI, interspecific competition intensity. * $p < 0.05$; ** $p < 0.01$.

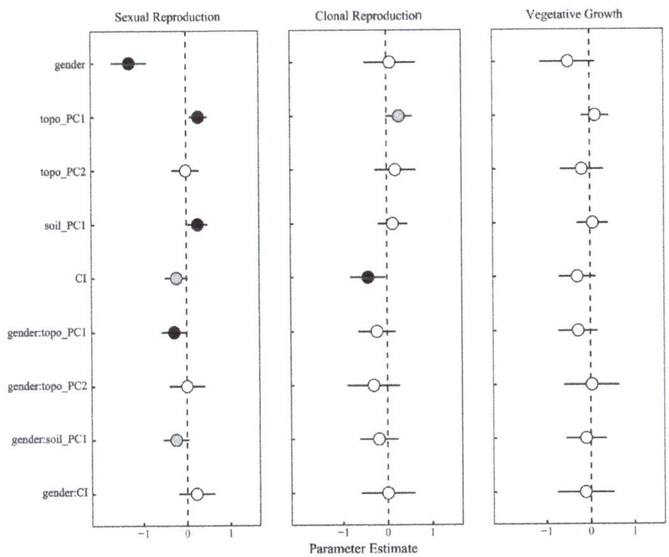


**Figure 2.** Effects of environmental variables on sexual reproduction, clonal reproduction, and vegetative growth. We show the parameter estimates of model predictors and the associated 95% confidence intervals. The *p*-values of each predictor are given as: Black indicates $p < 0.05$, grey indicates 0.05 $p < 0.1$, and white indicates $p > 0.1$. Gender, based on males; topo_PC1, the first axis of the five topography variables; topo_PC2, the second axis of the five topography variables; soil_PC1, the first axis of the four soil nutrients variables; CI, interspecific competition; gender:topo_PC1, the interaction between gender and the first axis of the five topography variables; gender:topo_PC2, the interaction between gender and the second axis of the five topography variables; gender:soil_PC1, the interaction between gender and the first axis of the four soil nutrients variables; gender:CI, the interaction between gender and interspecific competition.

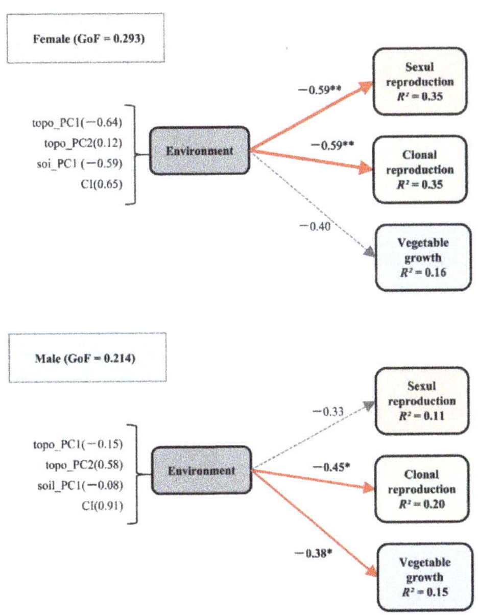


**Figure 3.** Structural equation model about the impact of the environment on sexual reproduction, clonal reproduction, and vegetative growth. Each number in parentheses indicates the loading value

of the indicator to the latent variable. Red lines indicate negative significant relationships, respectively, and gray lines indicate insignificant relationships; the thickness of the line represents the strength of the causal relationship, supplemented by a standardized path coefficient. $R^2$ indicates the total variation of a dependent variable as explained by independent variables; GOF indicates the goodness of fit of the entire model. * $p < 0.05$; ** $p < 0.01$.

### 2.2. The Effects of Biotic and Abiotic Factors on Clonal Reproduction

In the Spearman's correlation analysis, clonal reproduction in females is significantly positively correlated with slope and cosA, and significantly negatively correlated with interspecific competition intensity. However, clonal reproduction in males is only significantly negatively correlated with interspecific competition intensity (Figure 1). The multiple regression model for clonal reproduction (Figure 2) reveals that gender does not have a significant effect on rematnumber ($p = 0.80$). However, CI exhibits a significant negative effect on rematnumber ($\beta = -0.43$, $p < 0.05$). Moreover, no significant interaction effects between the variables and gender were observed for number of ramets. In the PLS-PM models (Figure 3), the environment has a significant negative effect on female clonal reproduction ($\beta = -0.60$, $p < 0.01$), and it also exerts a significant negative influence on male sexual reproduction ($\beta = -0.45$, $p < 0.05$).

### 2.3. The Effects of Biotic and Abiotic Factors on Vegetative Growth

In the Spearman's correlation analysis, vegetative growth in females does not show significant correlations with environmental factors, while vegetative growth in males is significantly negatively correlated only with slope (Figure 1). The multiple regression model for vegetative growth (Figure 2) indicates that gender does not have a significant effect on plant biomass ($p = 0.12$). The independent variables, including CI, topo_PC1, topo_PC2, and soil_PC1, also do not exhibit significant effects on plant biomass. Furthermore, no significant interaction effects, such as those between gender and the independent variables, were observed. In the PLS-PM models (Figure 3), the path coefficient for the environmental factor on female vegetative growth is −0.400 with a $p$-value of 0.08, which is close to the significance threshold of 0.05, suggesting a potential negative impact of the environment on vegetative growth, though this effect is not fully significant. In contrast, the environment exerts a significant negative effect on male vegetative growth ($\beta = -0.38$, $p < 0.05$).

## 3. Discussion

Female and male plants experience different selective pressures and have different evolutionary directions. Their evolutionary differences can induce a range of morphological, physiological, and ecological differences between males and females, which are referred to as sexual dimorphism. Sexual dimorphism is not fixed under different environmental pressures, and males and females differ in their plastic responses to environmental factors [35]. In our study, although principal component analyses showed no significant differences between female and male *Acer barbinerve* on environmental factors such as topography, soil nutrients, and interspecific competition intensity, this result suggests that overall variation in these environmental factors does not directly distinguish between the sexes in terms of growth and reproduction on a macro scale (Figure 4). However, subsequent analyses revealed significant sexual differences in the response to these environmental factors. This indicates that, despite the overall similarity of environmental factors, male and female *Acer barbinerve* exhibit sex-specific responses to growth and reproduction.

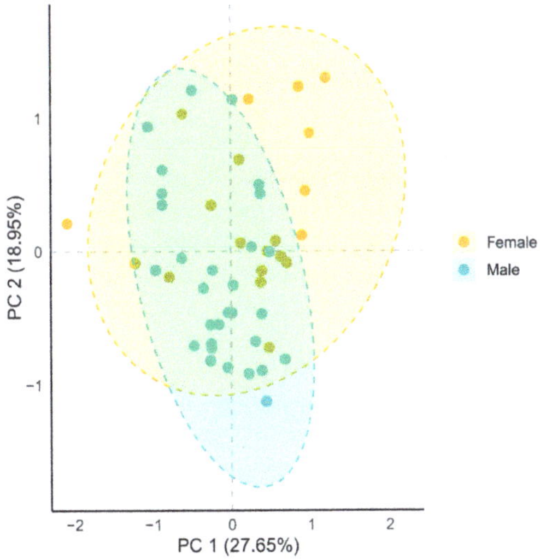


**Figure 4.** Principal component analysis of the topography, soil nutrients, and interspecific competition intensity of *Acer barbinerve* genets.

### *3.1. Differences in the Response of Sexual Reproduction to Environmental Factors Between the Sexes*

This study reveals that sexual reproduction in females and males responds differently to environmental factors. Topography and soil nutrients have a significant impact on sexual reproduction, with a notable interaction between topography and gender. Topographic factors are critical in creating habitat differences across regions [36]. At small scales, elevation and slope are two of the most important topographic factors influencing plant functional traits [23]. In the response of three grassland plants to changes in elevation, it was found that the reproduction of *Trifolium montanum* and *Briza media* was almost unaffected by elevation, whereas *Ranunculus bulbosus* showed a significant increase in reproductive investment at higher elevations [37]. Olejniczak et al. also demonstrated a species-specific dependence of seed production on changes in elevation [38]. In this study, the interaction between sex and topo_PC1 suggests a weak effect of topographic factors on males This may be related to differences in water demand or water use efficiency between the sexes [39–41]. In addition to elevation and slope, aspect also affects sexual reproduction in plants. For the dioecious shrub *Bursera fagaroides* (Burseraceae), reproduction was found to be related to aspect, with individuals growing on west-facing slopes producing lower fruit production than those growing on east-facing slopes [42,43].

Male and female plants face different resource limitations during sexual reproduction, with female functions (i.e., seed and fruit production) generally requiring more carbon [25,44], while male functions may have higher nitrogen demands for pollen production [45,46], which in turn leads to differences in their nutrient requirements and resource allocation. Different resource currencies (e.g., carbon vs. nitrogen) in dioecious plants may be an important factor driving sex-specific patterns of sexual dimorphism [47]. Sexual reproduction in males is limited to flowering and pollen dispersal, while females continue with fertilization, seed, and fruit production and dispersal [48]. This study found that the sexual reproduction of males was significantly enhanced with increasing soil nutrients. The increase in soil nutrient content enhanced nitrogen uptake by male flowers, directing more nitrogen toward pollen production, which subsequently increased male flower production. This, in turn, improved the fertilization success rate of female flowers, ultimately leading to a higher number of successfully developed samaras.

In a study of the dioecious wind-pollinated shrub *Leucadendron rubrum*, it was found that plant density negatively impacted the male fitness proxy in males but not significantly in females [49]. Similarly, our study found that sexual reproduction in males was significantly suppressed by the intensity of interspecific competition, whereas females were unaffected. The negative effect of interspecific competition on male sexual reproduction may be due to the increased competition for nutrient resources that affect pollen production. Females showed a stronger sexual reproduction in high competitive environments, possibly because female plants prioritize investing more resources in seed development to ensure the reproductive success of their offspring [27,50].

These results suggest that although male and female *Acer barbinerve* grow in similar environments, their sexual reproduction varies in response to environmental factors. These sex-specific response patterns reflect differences in reproductive strategies and reveal the ecological adaptation strategies of different sexes in varying environments.

### *3.2. Differences in the Response of Clonal Reproduction to Environmental Factors Between the Sexes*

Plants have repeatedly evolved clonal reproduction alongside sexual reproduction [51], forming an integrated strategy designed to ensure the transmission of adaptive genes while also providing the genetic variability necessary for colonizing new habitats and surviving future environmental changes [52]. In nature, the availability of basic resources (e.g., light, water, and soil nutrients) typically exhibits spatial heterogeneity.

Clonal plant ramets are often physically connected for extended periods through horizontal connectors (stolons and rhizomes), allowing for physiological integration between ramets [53,54]. Due to clonal integration, resources obtained by ramets growing in favorable microhabitats can be transported to resource-scarce locations, alleviating resource shortages in impoverished microsites [55]. Clonal plants can exhibit foraging responses, in which more resource-absorbing organs (e.g., leaves, roots, or ramets) are placed in high-quality patches than in low-quality ones, enabling the efficient use of heterogeneously distributed light and water resources [56,57]. Such cooperative systems can buffer the effects of spatial heterogeneity [58], and improve the overall performance of the plant [32,59].

It is due to these characteristics of clonal plants that topography and soil nutrient content are not the major factors influencing the clonal reproduction of *Acer barbinerve* under natural field conditions. Similarly to our study, when examining the reproductive allocation of dioecious aquatic clonal plants in response to water depth, it was found that both sexes invested similarly in clonal reproduction (via tubers) [47]. In general, the importance of clonal reproduction tends to increase with elevation. At higher elevations (3944 m), *Fragaria vesca* showed stronger clonal integration effects in terms of biomass, total stolon length, and number of ramets [60]. In a study on the variation of sexual and clonal reproduction in the alpine pioneer plant *Geum reptans* at different elevations, it was found that clonal reproduction was more common in low- and high-elevation populations, but less frequent in mid-elevation populations [61]. At the same time, the proportion of sexual and clonal reproduction in *Geum reptans* was more determined by the plastic responses of individual plants to local environmental conditions, rather than being solely driven by environmental gradients (e.g., elevation, etc.) [61]. This further emphasizes the importance of integrative studies.

In addition to resource heterogeneity, clonal plants are affected by many other environmental stressors, of which interspecific competition is an important factor. Studies have shown that survival, clonal reproduction, and most growth traits of clonal plants are reduced by competition [62]. Furthermore, the reproductive output of all plants (whether clonal or not) increases with improved habitat nutrient status and light conditions, while it declines under unfavorable moisture conditions [63]. The proportion of clonal plants usually varies across habitat gradients, which may be due to the fact that sexual reproduction becomes more difficult under certain environmental conditions, and clonal plants are able to bypass the barriers to sexual reproduction through clonal reproduction, thus gaining an adaptive advantage [64]. In clonal dioecious plants, resource allocation patterns may differ between the sexes: males typically allocate more resources to clonal structures, while females allocate less [3,65]. However, in the absence of competition, male and female plants of *Antennaria dioica* do not show significant differences in resource allocation [66]. In this study, clonal reproduction in both female and male *Acer barbinerve* was significantly negatively impacted by the intensity of interspecific competition, possibly because there was no significant difference in resource allocation for clonal reproduction between the sexes. Similarly, in the presence of *Phragmites australis*, the ramet density and rhizome sprouting rate of *Typha* spp. significantly decreased, indicating that interspecific competition negatively affected the clonal reproduction of *Typha* spp. [31].

### 3.3. Differences in the Response of Vegetative Growth to Environmental Factors Between the Sexes

Both biotic and abiotic factors affect biomass accumulation [67]. Competition alters the availability of nutrients, water, and light, thereby affecting plant growth and community structure in a given environment [9]. In this study, the vegetative growth of males was significantly negatively affected by environmental factors. Similarly to our study, Eppley found that females were more competitive than males in growth at high densities [68]. The presence of competition reduces plant biomass [69]. In studies on the impact of *Phragmites australis* invasion on *Typha* spp., it was found that the underground biomass of both plants decreased [31]. Consistent with previous studies, in this research, as the intensity of interspecific competition increased, the nutrient biomass of both sexes showed a declining trend.

Plants exhibit local adaptation through adaptive plasticity, altering their morphological or physiological traits along environmental gradients [70]. A recent study has shown that plant biomass follows a unimodal relationship with elevation, with resources concentrated in mid-elevation areas [71,72]. No significant relationship was found between the nutrient biomass of female and male plants with elevation in this study, probably because of the narrow range of elevation (435.47–501.96 m), which was not enough to show a significant elevation gradient effect, or because the environmental changes in this range had a more limited effect on plant growth and failed to cause a significant difference in biomass. Slope and aspect affect the amount of solar radiation received by vegetation surfaces, and solar radiation is a major component of surface energy balance, directly influencing microclimate conditions (e.g., near-surface temperature, evaporative demand, soil moisture, etc.) [20]. In mid-latitude regions, aspect has the most significant effect on ecological processes and soil properties [16,19]. North-facing slopes (shady slopes) typically have thicker, denser vegetation and nutrient-rich soils, while south-facing slopes (sunny slopes) have sparser vegetation, poorer soil development, and higher erosion rates. On south-facing slopes, species with a higher leaf mass area ratio (LMA) and smaller leaves are dominant, while on north-facing slopes, species with a lower LMA and larger leaves dominate [73]. Topography has a significant effect on tree growth, with trees growing faster in valleys than on ridges [74]. In this study, soil nutrient content had a relatively low impact on the vegetative growth of both female and male *Acer barbinerve*, indicating that the soil nutrient content in this area is generally sufficient to meet the vegetative growth needs of *Acer barbinerve*, and is not a limiting factor for vegetative growth.

## 4. Materials and Methods

### 4.1. Overview of the Study Area

This study was carried out in a natural temperate forest located in the Jiaohe Management Bureau of the Experimental Forest Zone in Jilin Province, northeastern China (43.97° N, 127.71–127.72° E). The area is characterized by a temperate continental climate, influenced by the monsoon, with an average annual temperature of 3.8 °C and an annual precipitation of 695.9 mm. The average monthly temperature ranges from −18.6 °C (January) to 21.7 °C (July). The soil type is mountainous dark brown forest soil with a thickness of 20–100 cm. In July 2009, a permanent forest monitoring study was initiated in the area, covering an area of 11.52 $hm^2$ (320 m × 360 m) [75]. The study area is subdivided into 272 contiguous plots, each measuring 20 m × 20 m. All woody plant individuals with diameters at breast height (DBH) $\geq$ 1 cm were identified, labeled, measured, and located at the species level. The first survey of the study area was completed in 2009, and re-measurements were conducted every five years using the same methodology. The forest stand is classified as near-mature, with the dominant tree species in the main canopy including *Fraxinus mandshurica, Tilia amurensis, Pinus koraiensis, Quercus mongolica, Acer pictum* subsp. *mono, Ulmus davidiana* var. *japonica, Carpinus cordata*, and *Juglans mandshurica*. The understory is mainly composed of *Syringa reticulata* subsp. *amurensis, Corylus mandshurica*, and *Acer barbinerve* [33].

### 4.2. Survey Sampling

During the flowering period (May 2023), female and male *Acer barbinerve* genets were randomly selected in the study area. Each genet consisted of several ramets, and the basal survey was carried out on each ramet of the genet. During the flowering period, the numbers of female and male flowers were counted, and the corresponding floral biomass was calculated. During the fruiting period (July 2023), a follow-up basic survey was conducted on the selected genets. The number of ramets, number of leaves, length of new branches, and number of samaras (only for female plants) were recorded, and the corresponding biomass was calculated. Genets that died during the survey period were excluded. The final number of surviving genets at the fruiting period was 20 female plants and 30 male plants. The basic distribution of the genets is shown in Table S1.

For each ramet, 5–10 representative branches were marked at different orientations to reflect the average condition of the ramet. During the flowering and fruiting periods, the number of flowers, fruits, and leaves, and the length of new branches on the marked branches and the total number of branches of the ramets were counted. From these, the total numbers of flowers, fruits, and leaves, and the total new branch length for the plant were calculated. Corresponding samples of flowers, fruits, leaves, and new branches were collected. The flower, fruit, and leaf samples were dried to a constant weight at 65 °C, and then weighed using an analytical balance (AUW220D, Shimadzu, Shanghai, China) to determine their dry biomass. This allowed for the calculation of the biomass per flower, fruit, and leaf, and the biomass of new branches per unit length. By multiplying these values with the plant's total numbers of flowers, fruits, leaves, and total new branch length, the total flower, fruit, and leaf biomass and new branch biomass of the plant were obtained.

The floral biomass and fruit biomass together constitute the sexual reproduction biomass, while the biomass of new branches and leaves together constitute the vegetative biomass. In this study, sexual reproduction biomass was used as the indicator for sexual reproduction, the number of ramets as the indicator for clonal reproduction, and vegetative biomass as the indicator for vegetative growth.

### 4.3. Abiotic Factors

#### 4.3.1. Topography

Topography variables include elevation, slope, aspect, and convexity. Elevation data for the plot were recorded using a GPS at the four corners of the plot, and the average value was then calculated. Slope was determined by calculating the average angular deviation between each angle of the four triangular planes formed by connecting three of the four corners and the horizontal plane, which provides the slope value for the plot. Aspect refers to the direction that the slope faces, with the direction due south defined as 0°, and due west, due north, and due east defined as 90°, 180°, and 270°, respectively, in a clockwise direction. Since aspect is a circular variable, we standardized it using a method [76] to reflect the northerly (cosA) and easterly (sinA) orientations of the slope (ranging from −1 to +1). These two variables increase with more northerly and easterly aspects, respectively. Convexity is defined as the difference between the elevation of the center of the plot and the average elevation of all adjacent plots.

#### 4.3.2. Soil Nutrients

In the four directions (east, south, west, and north) of the selected *Acer barbinerve* genets, avoiding plant roots and rocks, 100 g of soil from the 0–10 cm soil layer was collected using an iron shovel from each of the four directions. The soil from all four directions was placed in cloth bags and mixed into a single composite soil sample. The sample was air-dried and then passed through 2 mm and 0.15 mm sieves. The powder was compressed into thin disks (pressure: 6 t) using a benchtop powder pelletizer, placed in plastic-sealed bags, and labeled. Finally, the carbon, nitrogen, and phosphorus content of the soil (in units of $mg \cdot kg^{-1}$) was measured using the J200 LA-LIBS Laser Spectroscopic Element Analyzer (Applied Spectra, Fremont, CA, USA). To avoid interference between male and female plants, samples from males and females were measured separately.

### 4.4. Biotic Factors

We used interspecific competition intensity (CI) to represent biotic factors, as it reflects, to some extent, the species richness around the genets and the size of surrounding genets, which can effectively show the influence of surrounding species on the research subject. In this study, the Hegyi index [77] is used to quantify the interspecific competition intensity. The diameter at breast height (DBH) of *Acer barbinerve* genets is used to represent tree size, as shown in the following formula:

$$CI = \sum_{j=1}^{m} \frac{D_j}{D_i} \times \frac{1}{L_{ij}}$$

where $D_i$ is the DBH of *Acer barbinerve* genets, $D_j$ is the DBH of a competitor tree, $L_{ij}$ is the distance between the competitor tree and the *Acer barbinerve* genets, $m$ is the number of competitor trees. A larger CI value indicates greater competition pressure on *Acer barbinerve* genets. Based on the coordinates of the trees in the plot, the distance between the competitor trees and the *Acer barbinerve* genets can be calculated. We use the fixed circular plot method to calculate the competition intensity of genets. According to previous studies, the competition radius for both male and female *Acer barbinerve* genets is 4 m [28]. All other species within this radius are considered competitor trees.

### 4.5. Data Analysis

We first performed descriptive statistical analyses on the biotic and abiotic factors within the study area, and employed ANOVA to examine significant differences in these factors between sexes (Table S1). To assess whether significant differences existed in the overall environmental conditions between males and females, principal component analysis (PCA) was conducted. Spearman's correlation analysis was implemented to evaluate the nonparametric correlation strength between pairwise variables (Figure 2). Prior to analytical procedures, all predictor and response variables were standardized using Z-scores. All model predictors had a variance inflation factor (VIF) < 4 (Table S2), indicating that multicollinearity between predictor variables was not a concern. Given the unequal sample sizes between sexes that might influence statistical outcomes, we performed multiple regression analyses incorporating interaction terms between sex and topographic features, soil nutrients, and interspecific competition to evaluate the environmental impacts on biological individuals. For multivariate model construction, topographic variables were subjected to PCA, with the first two principal components (topo_PC1 and topo_PC2) (cumulatively explaining 67.4% of variance) being retained (Figure S1). Similarly, soil variables underwent PCA, where the first principal component (soil_PC1) (explaining 55.5% of variance) was selected (Figure S1). This dimensionality reduction preserved essential information while enhancing regression stability. To further investigate the differential effects of environmental factors (including topo_PC1, topo_PC1, soil_PC1, and interspecific competition) on sexual reproduction, clonal reproduction, and vegetative growth, we constructed grouped partial least squares path modeling (PLS-PM) using the "plspm" package. In this framework, sex served as the grouping variable, enabling path analysis to discern sex-specific responses in growth and reproductive strategies under environmental influences. All statistical analyses were conducted in R 4.3.3.

## 5. Conclusions

Both abiotic and biotic factors significantly influence sexual reproduction, clonal reproduction, and vegetative growth in *Acer barbinerve*, with distinct responses between sexes. Environmental variables, particularly topography and soil nutrients, interact differently with males and females, affecting their reproductive strategies and growth patterns. While female sexual reproduction is more sensitive to environmental factors, interspecific competition negatively impacts both male and female clonal reproduction. These findings highlight the complex interplay between abiotic and biotic factors in shaping the growth and reproductive strategies of dioecious plants in diverse environments.

**Supplementary Materials:** The following supporting information can be downloaded at: https://www.mdpi.com/article/10.3390/plants14040596/s1, Table S1: Basic Information of the selected *Acer barbinerve* genets. Table S2: Effects of environmental variables on sexual reproduction, clonal reproduction, and vegetative growth. Figure S1: The biplot resulting from the principal component analysis for topography and soil nutrients.

**Author Contributions:** Conceptualization, J.W. and D.L.; methodology, D.L. and J.W.; software, D.L. and J.L.; validation, J.W.; formal analysis, D.L. and J.W.; investigation, D.L., J.L., Y.W. and Y.B.; data curation, D.L.; writing—original draft preparation, D.L.; writing—review and editing, D.L., X.Z. and J.W.; project administration, J.W.; funding acquisition, J.W. All authors have read and agreed to the published version of the manuscript.

**Funding:** This research was funded by the National Key R&D Program of China (2023YFF1304004-06).

**Data Availability Statement:** Data will be made available on reasonable request to the corresponding author.

**Conflicts of Interest:** The authors declare no conflicts of interest.

## References

1. Renner, S.S.; Ricklefs, R.E. Dioecy and Its Correlates in the Flowering Plants. *Am. J. Bot.* **1995**, *82*, 596–606. [CrossRef]
2. Darwin, C. *The Descent of Man and Selection in Relation to Sex*; John Murray: London, UK, 1871; Volume 1, ISBN 978-1-108-00509-8.

3. Barrett, S.C.H.; Hough, J. Sexual Dimorphism in Flowering Plants. *J. Exp. Bot.* **2013**, *64*, 67–82. [CrossRef] [PubMed]
4. Reznick, D. Costs of Reproduction: An Evaluation of the Empirical Evidence. *Oikos* **1985**, *44*, 257–267. [CrossRef]
5. Obeso, J.R. The Costs of Reproduction in Plants. *New Phytol.* **2002**, *155*, 321–348. [CrossRef]
6. Suzuki, N.; Rivero, R.M.; Shulaev, V.; Blumwald, E.; Mittler, R. Abiotic and Biotic Stress Combinations. *New Phytol.* **2014**, *203*, 32–43. [CrossRef]
7. Yu, L.; Huang, Z.; Tang, S.; Korpelainen, H.; Li, C. *Populus euphratica* Males Exhibit Stronger Drought and Salt Stress Resistance than Females. *Environ. Exp. Bot.* **2023**, *205*, 105114. [CrossRef]
8. Liu, G.; Wang, Y.; Lian, B.; Ma, Z.; Xiang, X.; Wu, J.; Luo, C.; Ma, D.; Chen, Y.; Yu, C.; et al. Molecular Responses to Salinity Stress in *Salix matsudana* (Koidz) Females and Males. *Front. Plant Sci.* **2023**, *14*, 1122197. [CrossRef]
9. Chen, J.; Dong, T.; Duan, B.; Korpelainen, H.; Niinemets, Ü.; Li, C. Sexual Competition and N Supply Interactively Affect the Dimorphism and Competiveness of Opposite Sexes in *Opulus cathayana*. *Plant Cell Environ.* **2015**, *38*, 1285–1298. [CrossRef]
10. Rabska, M.; Pers-Kamczyc, E.; Żytkowiak, R.; Adamczyk, D.; Iszkuło, G. Sexual Dimorphism in the Chemical Composition of Male and Female in the Dioecious Tree, *Juniperus communis* L., Growing under Different Nutritional Conditions. *Int. J. Mol. Sci.* **2020**, *21*, 8094. [CrossRef]
11. Bürli, S.; Pannell, J.R.; Tonnabel, J. Environmental Variation in Sex Ratios and Sexual Dimorphism in Three Wind-Pollinated Dioecious Plant Species. *Oikos* **2022**, *2022*, e08651. [CrossRef]
12. Juvany, M.; Munné-Bosch, S. Sex-Related Differences in Stress Tolerance in Dioecious Plants: A Critical Appraisal in a Physiological Context. *J. Exp. Bot.* **2015**, *66*, 6083–6092. [CrossRef] [PubMed]
13. Retuerto, R.; Sánchez Vilas, J.; Varga, S. Sexual Dimorphism in Response to Stress. *Environ. Exp. Bot.* **2018**, *146*, 1–4. [CrossRef]
14. Dudley, L.S. Ecological Correlates of Secondary Sexual Dimorphism in *Salix glauca* (Salicaceae). *Am. J. Bot.* **2006**, *93*, 1775–1783. [CrossRef] [PubMed]
15. Puixeu, G.; Pickup, M.; Field, D.L.; Barrett, S.C.H. Variation in Sexual Dimorphism in a Wind-Pollinated Plant: The Influence of Geographical Context and Life-Cycle Dynamics. *New Phytol.* **2019**, *224*, 1108–1120. [CrossRef]
16. Singh, S. Understanding the Role of Slope Aspect in Shaping the Vegetation Attributes and Soil Properties in Montane Ecosystems. *Trop. Ecol.* **2018**, *59*, 417–430.
17. Shen, Z.; Zhan, X.; Jin, Y. Gradient Analysis of the Influence of Mountain Topography on Vegetation Pattern. *Chin. J. Plant Ecol.* **2000**, *24*, 430–435.
18. Comita, L.S.; Engelbrecht, B.M.J. Seasonal and Spatial Variation in Water Availability Drive Habitat Associations in a Tropical Forest. *Ecology* **2009**, *90*, 2755–2765. [CrossRef]
19. Zhang, Q.; Fang, R.; Deng, C.; Zhao, H.; Shen, M.-H.; Wang, Q. Slope Aspect Effects on Plant Community Characteristics and Soil Properties of Alpine Meadows on Eastern Qinghai-Tibetan Plateau. *Ecol. Indic.* **2022**, *143*, 109400. [CrossRef]
20. Bennie, J.; Huntley, B.; Wiltshire, A.; Hill, M.O.; Baxter, R. Slope, Aspect and Climate: Spatially Explicit and Implicit Models of Topographic Microclimate in Chalk Grassland. *Ecol. Model.* **2008**, *216*, 47–59. [CrossRef]
21. Fan, J.; Xu, Y.; Ge, H.; Yang, W. Vegetation Growth Variation in Relation to Topography in Horqin Sandy Land. *Ecol. Indic.* **2020**, *113*, 106215. [CrossRef]
22. Sterner, R.W.; Elser, J.J. *Ecological Stoichiometry: The Biology of Elements from Molecules to the Biosphere*; Princeton University Press: Princeton, NJ, USA, 2003.
23. Yin, S. Carbon, Nitrogen and Phosphorus Contents and Their Ecological Stoichiometric Characteristics in Leaf Litter from the Jianfengling Tropical Montane Rainforest. *Front. Plant Sci.* **2024**, *15*, 1478094. [CrossRef]
24. Zhang, Q.; Yeerjiang, B.; Wang, J. Gender dimorphism in nutritional resource requirements of dioecious tree species *Rhamnus schneideri* var. *manshurica*. *Chin. J. Plant Ecol.* **2023**, *47*, 1708–1717. [CrossRef]
25. McDowell, S.C.; McDowell, N.G.; Marshall, J.D.; Hultine, K. Carbon and Nitrogen Allocation to Male and Female Reproduction in Rocky Mountain Douglas-Fir (*Pseudotsuga menziesii* var. *glauca*, Pinaceae). *Am. J. Bot.* **2000**, *87*, 539–546. [CrossRef] [PubMed]
26. De Lisle, S.P. Rapid Evolution of Ecological Sexual Dimorphism Driven by Resource Competition. *Ecol. Lett.* **2023**, *26*, 124–131. [CrossRef] [PubMed]
27. Tonnabel, J.; David, P.; Pannell, J.R. Sex-specific Strategies of Resource Allocation in Response to Competition for Light in a Dioecious Plant. *Oecologia* **2017**, *185*, 675–686. [CrossRef] [PubMed]
28. Liao, D.; Wang, Y.; Lei, J.; Wang, Y.; Wang, J. Gender-specific response to interspecific competition in dioecious clonal plant *Acer barbinerve*. *Chin. J. Plant Ecol.* **2024**, *48*, 1623–1636. [CrossRef]
29. Bennett, J.A.; Riibak, K.; Tamme, R.; Lewis, R.J.; Pärtel, M. The Reciprocal Relationship between Competition and Intraspecific Trait Variation. *J. Ecol.* **2016**, *104*, 1410–1420. [CrossRef]
30. Yu, L.; Song, M.; Lei, Y.; Korpelainen, H.; Niinemets, Ü.; Li, C. Effects of Competition and Phosphorus Fertilization on Leaf and Root Traits of Late-Successional Conifers *Abies fabri* and *Picea brachytyla*. *Environ. Exp. Bot.* **2019**, *162*, 14–24. [CrossRef]
31. Chun, Y.-M.; Choi, Y.D. Expansion of *Phragmites australis* (Cav.) Trin. Ex Steud. (Common Reed) into *Typha* Spp. (Cattail) Wetlands in Northwestern Indiana, USA. *J. Plant Biol.* **2009**, *52*, 220–228. [CrossRef]

32. Wang, Y.-J.; Shi, X.-P.; Meng, X.-F.; Wu, X.-J.; Luo, F.-L.; Yu, F.-H. Effects of Spatial Patch Arrangement and Scale of Covarying Resources on Growth and Intraspecific Competition of a Clonal Plant. *Front. Plant Sci.* **2016**, *7*, 753. [CrossRef]
33. Qiao, X.; Zhang, J.; Wang, Z.; Xu, Y.; Zhou, T.; Mi, X.; Cao, M.; Ye, W.; Jin, G.; Hao, Z.; et al. Foundation Species across a Latitudinal Gradient in China. *Ecology* **2021**, *102*, e03234. [CrossRef] [PubMed]
34. Pan, C.; Zhang, C.; Zhao, X.; Xia, F.; Zhou, H.; Wang, Y. Sex Ratio and Spatial Patterns of Males and Females of Different Ages in the Dioecious Understory Tree, *Acer barbinerve*, in a Broad-Leaved Korean Pine Forest. *Biodivers. Sci.* **2010**, *18*, 292–299. [CrossRef]
35. Delph, L.F.; Bell, D.L. A Test of the Differential-Plasticity Hypothesis for Variation in the Degree of Sexual Dimorphism in *Silene latifolia*. *Evol. Ecol. Res.* **2008**, *10*, 61–75.
36. Cao, K.; Rao, M.; Yu, J.; Liu, X.; Mi, X.; Chen, J. The phylogenetic signal of functional traits and their effects on community structure in an evergreen broad-leaved forest. *Biodivers. Sci.* **2013**, *21*, 564. [CrossRef]
37. Frei, E.R.; Ghazoul, J.; Matter, P.; Heggli, M.; Pluess, A.R. Plant Population Differentiation and Climate Change: Responses of Grassland Species along an Elevational Gradient. *Glob. Change Biol.* **2014**, *20*, 441–455. [CrossRef]
38. Olejniczak, P.; Czarnoleski, M.; Delimat, A.; Majcher, B.M.; Szczepka, K. Seed Size in Mountain Herbaceous Plants Changes with Elevation in a Species-Specific Manner. *PLoS ONE* **2018**, *13*, e0199224. [CrossRef]
39. Sánchez-Vilas, J.; Bermúdez, R.; Retuerto, R. Soil Water Content and Patterns of Allocation to Below- and above-Ground Biomass in the Sexes of the Subdioecious Plant *Honckenya peploides*. *Ann. Bot.* **2012**, *110*, 839–848. [CrossRef]
40. Zhang, X.; Zhang, C.; Zhao, X. Effect of Sex Ratio, Habitat Factors and Neighborhood Competition on Stem Growth in the Dioecious Tree *Fraxinus mandshurica*. *Ecol. Res.* **2014**, *29*, 309–317. [CrossRef]
41. Hultine, K.R.; Grady, K.C.; Wood, T.E.; Shuster, S.M.; Stella, J.C.; Whitham, T.G. Climate Change Perils for Dioecious Plant Species. *Nat. Plants* **2016**, *2*, 1–8. [CrossRef]
42. Ortiz-Pulido, R.; Rico-Gray, V. Seed Dispersal of *Bursera fagaroides* (Burseraceae): The Effect of Linking Environmental Factors. *Southwest. Nat.* **2006**, *51*, 11–21. [CrossRef]
43. Ortiz-Pulido, R.; Pavón, N.P. Influence of Slope Orientation on Sex Ratio and Size Distribution in a Dioecious Plant *Bursera fagaroides* Var. *Purpusii* (Brandeg.) McVaugh and Rzed. (Burseraceae). *Plant Ecol.* **2010**, *208*, 271–277. [CrossRef]
44. Antos, J.; Allen, G. A Comparison of Reproductive Effort in the Dioecious Shrub *Oemleria cerasiformis* Using Nitrogen, Energy and Biomass as Currencies. *Am. Midl. Nat.* **1990**, *124*, 254. [CrossRef]
45. Harris, M.S.; Pannell, J.R. Roots, Shoots and Reproduction: Sexual Dimorphism in Size and Costs of Reproductive Allocation in an Annual Herb. *Proc. R. Soc. B Biol. Sci.* **2008**, *275*, 2595–2602. [CrossRef] [PubMed]
46. Van Drunen, W.E.; Dorken, M.E. Trade-Offs between Clonal and Sexual Reproduction in *Sagittaria latifolia* (Alismataceae) Scale up to Affect the Fitness of Entire Clones. *New Phytol.* **2012**, *196*, 606–616. [CrossRef]
47. Li, L.; Barrett, S.C.H.; Song, Z.; Chen, J. Sex-Specific Plasticity of Reproductive Allocation in Response to Water Depth in a Clonal, Dioecious Macrophyte. *Am. J. Bot.* **2019**, *106*, 42–50. [CrossRef]
48. Lloyd, D.G.; Webb, C.J. Secondary Sex Characters in Plants. *Bot. Rev.* **1977**, *43*, 177–216. [CrossRef]
49. Tonnabel, J.; Klein, E.K.; Ronce, O.; Oddou-Muratorio, S.; Rousset, F.; Olivieri, I.; Courtiol, A.; Mignot, A. Sex-Specific Spatial Variation in Fitness in the Highly Dimorphic *Leucadendron rubrum*. *Mol. Ecol.* **2021**, *30*, 1721–1735. [CrossRef]
50. Hesse, E.; Pannell, J.R. Sexual Dimorphism in a Dioecious Population of the Wind-Pollinated Herb *Mercurialis annua*: The Interactive Effects of Resource Availability and Competition. *Ann. Bot.* **2011**, *107*, 1039–1045. [CrossRef]
51. Klimeš, L.; Klimešová, J.; Hendriks, R.; van Groenendael, J. Clonal Plant Architecture: A Comparative Analysis of Form and Function. In *The Ecology and Evolution of Clonal Plants*; de Kroon, H., van Groenendael, J., Eds.; Backhuys Publishers: Leiden, The Netherlands, 1997; pp. 1–29.
52. Niklas, K.J.; Cobb, E.D. The Evolutionary Ecology (Evo-Eco) of Plant Asexual Reproduction. *Evol. Ecol.* **2017**, *31*, 317–332. [CrossRef]
53. Dong, M.; Yu, F.; Chen, Y.; Song, M.; Liu, J.; Li, J.; Liu, F. *Ecology of Clonal Plants*; Science Press: Beijing, China, 2011; ISBN 978-7-03-030944-0.
54. Liu, F.; Liu, J.; Dong, M. Ecological Consequences of Clonal Integration in Plants. *Front. Plant Sci.* **2016**, *7*, 770. [CrossRef]
55. Hutchings, M.J.; Wijesinghe, D.K. Patchy Habitats, Division of Labour and Growth Dividends in Clonal Plants. *Trends Ecol. Evol.* **1997**, *12*, 390–394. [CrossRef] [PubMed]
56. De Kroon, H.; Huber, H.; Stuefer, J.F.; Van Groenendael, J.M. A Modular Concept of Phenotypic Plasticity in Plants. *New Phytol.* **2005**, *166*, 73–82. [CrossRef] [PubMed]
57. Guo, W.; Song, Y.-B.; Yu, F.-H. Heterogeneous Light Supply Affects Growth and Biomass Allocation of the Understory Fern *Diplopterygium glaucum* at High Patch Contrast. *PLoS ONE* **2011**, *6*, e27998. [CrossRef] [PubMed]
58. Roiloa, S.R.; Alpert, P.; Tharayil, N.; Hancock, G.; Bhowmik, P.C. Greater Capacity for Division of Labour in Clones of *Fragaria chiloensis* from Patchier Habitats. *J. Ecol.* **2007**, *95*, 397–405. [CrossRef]

59. Song, Y.-B.; Yu, F.-H.; Keser, L.H.; Dawson, W.; Fischer, M.; Dong, M.; van Kleunen, M. United We Stand, Divided We Fall: A Meta-Analysis of Experiments on Clonal Integration and Its Relationship to Invasiveness. *Oecologia* **2013**, *171*, 317–327. [CrossRef]
60. Chen, J.S.; Yu, D.; Liu, Q.; Dong, M. Clonal Integration of the Stoloniferous Herb *Fragaria vesca* from Different Altitudes in Southwest China. *Flora* **2004**, *199*, 342–350. [CrossRef]
61. Weppler, T.; Stöcklin, J. Variation of Sexual and Clonal Reproduction in the Alpine *Geum reptans* in Contrasting Altitudes and Successional Stages. *Basic Appl. Ecol.* **2005**, *6*, 305–316. [CrossRef]
62. Koutecka, E.; Leps, J. Performance of Three Closely Related Myosotis Species in an Experiment in Which Substrate Quality and Competition Were Manipulated. *Preslia* **2011**, *83*, 403–420.
63. Gomes, A.S.; Callaway, R.M.; Rabelo, B.S.; Petry, G.L.; Barbosa, E.M.; Borghetti, F. Competition for Water and Rapid Exclusion of an Island Endemic by a Pantropical Species in a Tropical Climate. *Oecologia* **2023**, *201*, 901–914. [CrossRef]
64. Herben, T.; Šerá, B.; Klimešová, J. Clonal Growth and Sexual Reproduction: Tradeoffs and Environmental Constraints. *Oikos* **2015**, *124*, 469–476. [CrossRef]
65. García, M.B.; Antor, R.J. Sex Ratio and Sexual Dimorphism in the Dioecious *Borderea pyrenaica* (Dioscoreaceae). *Oecologia* **1995**, *101*, 59–67. [CrossRef] [PubMed]
66. Varga, S.; Kytöviita, M.-M. Differential Competitive Ability between Sexes in the Dioecious *Antennaria dioica* (Asteraceae). *Ann. Bot.* **2012**, *110*, 1461–1470. [CrossRef]
67. Reich, P.B.; Luo, Y.; Bradford, J.B.; Poorter, H.; Perry, C.H.; Oleksyn, J. Temperature Drives Global Patterns in Forest Biomass Distribution in Leaves, Stems, and Roots. *Proc. Natl. Acad. Sci. USA* **2014**, *111*, 13721–13726. [CrossRef] [PubMed]
68. Eppley, S.M. Females Make Tough Neighbors: Sex-Specific Competitive Effects in Seedlings of a Dioecious Grass. *Oecologia* **2006**, *146*, 549–554. [CrossRef]
69. Varga, S.; Vega-Frutis, R.; Kytoviita, M.-M. Competitive Interactions Are Mediated in a Sex-Specific Manner by Arbuscular Mycorrhiza in *Antennaria dioica*. *Plant Biol.* **2017**, *19*, 217–226. [CrossRef]
70. Hirano, M.; Sakaguchi, S.; Takahashi, K. Phenotypic Differentiation of the *Solidago virgaurea* Complex along an Elevational Gradient: Insights from a Common Garden Experiment and Population Genetics. *Ecol. Evol.* **2017**, *7*, 6949–6962. [CrossRef]
71. Lee, M.A.; Burger, G.; Green, E.R.; Kooij, P.W. Relationships between Resource Availability and Elevation Vary between Metrics Creating Gradients of Nutritional Complexity. *Oecologia* **2021**, *195*, 213–223. [CrossRef]
72. Thakur, G.; Kumar, P.; Bhardwaj, D.R.; Prakash, P. Poonam Dynamics of Aboveground Vegetation Biomass and Carbon Stocks along the Altitudinal Gradients and Overstorey Composition Types in the Temperate Himalayan Region. *Trees For. People* **2024**, *16*, 100553. [CrossRef]
73. Li, X.; Song, X.; Zhao, J.; Lu, H.; Qian, C.; Zhao, X. Shifts and Plasticity of Plant Leaf Mass per Area and Leaf Size among Slope Aspects in a Subalpine Meadow. *Ecol. Evol.* **2021**, *11*, 14042–14055. [CrossRef]
74. Fortunel, C.; Lasky, J.R.; Uriarte, M.; Valencia, R.; Wright, S.J.; Garwood, N.C.; Kraft, N.J.B. Topography and Neighborhood Crowding Can Interact to Shape Species Growth and Distribution in a Diverse Amazonian Forest. *Ecology* **2018**, *99*, 2272–2283. [CrossRef]
75. Zhang, C.; Zhao, Y.; Zhao, X.; Von Gadow, K. Species-Habitat Associations in a Northern Temperate Forest in China. *Silva Fenn.* **2012**, *46*, 501–519. [CrossRef]
76. Zhong, L.; Didham, R.K.; Liu, J.; Jin, Y.; Yu, M. Community Re-Assembly and Divergence of Woody Plant Traits in an Island–Mainland System after More than 50 Years of Regeneration. *Divers. Distrib.* **2021**, *27*, 1435–1448. [CrossRef]
77. Hegyi, F. A Simulation Model for Managing Jack-Pine Stands. In *Growth Models for Tree and Stand Simulation*; Royal College Research Notes; Fries, J., Ed.; Royal College of Forestry: Stockholm, Sweden, 1974; Volume 30, pp. 74–90.

*plants*

*Article*

# Identification and Chemical Control of Stem Canker Pathogen of *Idesia polycarpa*

**Jian Feng [1,2,3,4,†], Qiupeng Yuan [1,2,3,†], Xuzhong Chen [1,5], Lisha Fang [1,2,3], Tao Zhang [1,2,3], Zhen Liu [1,2,3], Yanmei Wang [1,2,3], Xiaodong Geng [1,2,3], Qifei Cai [1,2,3,*] and Zhi Li [1,2,5,*]**

1 College of Forestry, Henan Agricultural University, Zhengzhou 450046, China
2 National Forestry and Grassland Administration Key Laboratory for Central Plains Forest Resources Cultivation, Zhengzhou 450046, China
3 Henan Province Engineering Technology Research Center for Idesia, Zhengzhou 450046, China
4 Key Laboratory for Bio-Resource and Eco-Environment of Ministry of Education, College of Life Sciences, Sichuan University, Chengdu 610065, China
5 Zhongyuan Environmental Protection Development Co., Ltd., Zhengzhou 450046, China
* Correspondence: cai_qifei@henau.edu.cn (Q.C.); lizhi@henau.edu.cn (Z.L.); Tel.: +86-13001292522 (Q.C.); +86-17752559889 (Z.L.)
† These authors contributed equally to this work.

**Abstract**

*Idesia polycarpa* is an important woody oilseed tree crucial for ensuring China's grain and oil security. The expansion of *I. polycarpa* plantations has been accompanied by an increase in pests and diseases, with canker disease recently observed in two forests in Henan Province. Field surveys revealed a disease incidence of 70.12% among 328 surveyed trees, indicating a substantial threat to plantation health. The most virulent pathogen, strain SQ5, was identified as *Botryosphaeria dothidea* through molecular sequencing and morphological analyses. Strain SQ5 showed an optimum growth temperature of 25 °C and a mycelial lethal temperature of 60 °C. The pathogen thrives in acidic conditions and is promoted by light, with the ability to utilize various carbon and nitrogen sources. In vitro toxicity assessments identified four effective fungicides: 70% thiophanate-methyl (EC50 = 0.0169 μg/mL), 43% tebuconazole (EC50 = 0.0219 μg/mL), 20% octylamine acetate (EC50 = 0.0271 μg/mL), and 40% difenoconazole (EC50 = 0.0954 μg/mL). Field trials demonstrated that 43% tebuconazole (average efficacy = 35.29%) and 40% difenoconazole (average efficacy = 23.53%) exhibited superior control of *I. polycarpa* canker. This study represents the first systematic analysis of *I. polycarpa* canker and its control measures, laying a foundation for further research and field management strategies. Given the significance of *I. polycarpa* in Chinese forestry, this underscores the need for effective management strategies to sustain its productivity and mitigate risks associated with expanding plantations.

**Keywords:** *Idesia polycarpa*; stem canker; *Botryosphaeria dothidea*; wood pathogens; fungicide

## 1. Introduction

Canker is a prevalent plant disease with widespread detrimental effects, impacting broad-leaved trees, coniferous trees, and herbaceous plants [1–3]. In particular, forest cankers cause localized necrosis in the trunk and branch cortex [4]. In mild cases, the disease disrupts the transport of nutrients and water within the tree, thereby impeding its growth and development. Without intervention, the disease progresses, significantly inhibiting growth and eventually leading to tree mortality. Notably, in 1994, 3000 beech trees succumbed to canker in the Great Smoky Mountains National Park, USA [5]. In 2015, *Populus tremuloides* across six ecological zones in Alaska experienced a 70% mortality rate due to canker disease [6]. Since 2018, *Pinus armandii* in Jingbian Village, Zhouqu County, China, has been affected by canker disease, with incidence rates increasing annually and reaching up to 80% in local plots [7]. Forest canker has emerged as a significant secondary forest pest in China, posing a threat to the forestry industry's sustainable development [8].

*Idesia polycarpa*, a dioecious plant from the genus *Idesia* within the family Flacourtiaceae, stands out as an ideal tree species for ecological enhancement due to its high adaptability, drought and barrenness tolerance, and minimal climatic requirements [9]. Additionally, it serves as a high-quality woody oil species, with oil content reaching 37.01% in the fruit pulp, 21.49% in the seeds, and 30.00% in the dried fruits [10]. These values surpass those of competitors such as *Camellia oleifera* and *Olea europaea* in the current market [11]. *I. polycarpa* oil is distinguished by its high quality and richness in unsaturated fatty acids, particularly linoleic acid, which constitutes an impressive 73.00% [12], making it a premium edible oil. Currently, *I. polycarpa* has been included in the national reserve forest initiative for superior industrial tree species, and its oil is classified as a general food product by the state [13]. The *I. polycarpa* industry has entered a phase of rapid and efficient growth, with promising market prospects. In recent years, the state has consistently increased support for the woody oilseed industry, implementing a series of policies [14]. Large-scale cultivation has been carried out across various regions, with *I. polycarpa* plantations covering 3300 $km^2$ in Hubei Province, over 66.67 $km^2$ in Sichuan Province, more than 13,333 $km^2$ in Shaotong City, Yunnan Province, and 1000 $km^2$ in Luoyang City, Henan Province, as of spring 2021. With growing recognition of *I. polycarpa*'s value, its cultivation area is anticipated to expand further.

It is generally believed that *I. polycarpa* has strong adaptability and resistance to diseases and pests. However, with the development of the *I. polycarpa* industry and the gradual expansion of artificial plantation areas, various diseases have emerged under the influence of environmental and human factors, seriously affecting the economic value of *I. polycarpa* and hindering the industry's healthy development. In a study, Zhou et al. [15] identified multiple diseases across seven *I. polycarpa* plantations in Sichuan Province, with plaster disease and anthracnose being the most common and damaging. Zhu [16] confirmed the presence of six types of diseases, including damping-off, anthracnose, and gray patch disease, in major *I. polycarpa* cultivation areas in Zhejiang Province. Notably, a new disease symptom was observed in two *I. polycarpa* forests in Henan Province: circular brown spots with water-soaked edges on tree stems, resembling canker symptoms previously reported in other tree species. Currently, there are no systematic reports on this disease affecting *I. polycarpa*.

Given this context, the objective of this study was to determine the incidence pattern of canker in *I. polycarpa* through field surveys. Subsequently, the aim was to identify the pathogen causing *I. polycarpa* canker through tissue culture, morphological observation, and molecular techniques, providing a theoretical basis for its prevention and control. Additionally, a pharmaceutical test was conducted to screen effective fungicides, offering a scientific reference for the field control of *I. polycarpa* canker.

## 2. Results

### 2.1. Symptoms of *I. polycarpa* Canker

Between 2020 and 2021, field surveys were conducted on *I. polycarpa* at two sites (Figure 1), based on typical canker disease symptoms, including trunk-localized lesions, circular brown spots, a water-soaked appearance, and internal tissue decay (Figure 2).

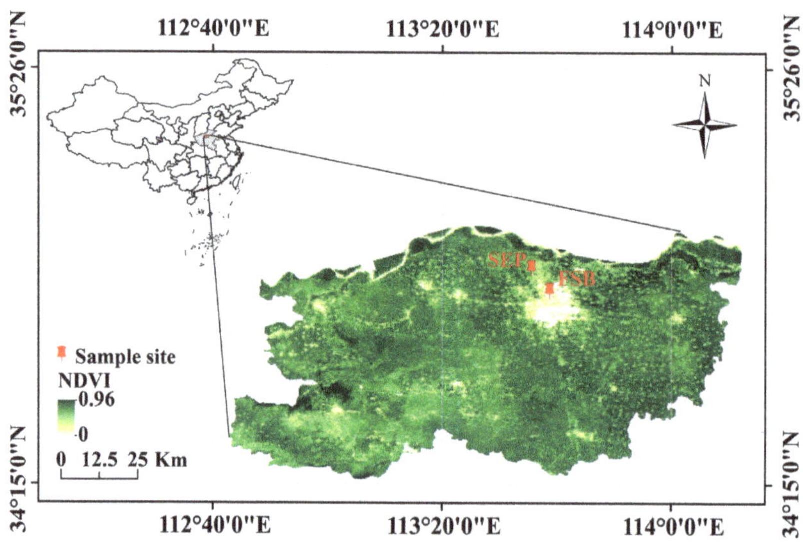


**Figure 1.** The spatial distribution of sampling sites in Zhengzhou, Henan Province, China. NDVI values were obtained from the MOD13Q1 product.

**Figure 2.** Pathogen isolation and inoculation. (**A**) Five isolated fungi of *Botryosphaeria dothidea*. (**B**) Inoculation of *I. polycarpa* seedlings.

A total of 328 trees were assessed, of which approximately 230 showed typical canker symptoms, and 17 symptomatic individuals died during the observation period. Disease incidence was 70.12%, calculated as the proportion of symptomatic trees among all surveyed individuals. These results indicate that canker disease poses a significant threat to the growth and survival of *I. polycarpa*. The disease typically initiates in mid to late March and persists until mid to late September. Considering the climate change patterns in Zhengzhou City (Figure S1), it can be concluded that its onset is concentrated in the hot and rainy seasons. Canker disease predominantly affects trees over 5 years old, with susceptible parts concentrated in the trunk.. As the disease progresses, parts of the branches and trunk will also appear as cankers. In the early stages, a sub-circular brown spot appears on the affected stem area, releasing brown fluid upon pressing, indicating internal decay (Figure 3A). Progressing further, the spots extend along the trunk into the xylem, forming patches that release significant amounts of brown water (Figure 3B,C). In later stages, the spots become sunken and the epidermis cracks, causing branches or the entire plant to die (Figure 3D). Additionally, we noted that on sunny days following rain, the severity of *I. polycarpa* canker increases, as observed after several days of intense heat. This pattern aligns with the overall morbidity trend observed in *I. polycarpa*.

**Figure 3.** Field symptoms of *Idesia. polycarpa* canker disease. (**A–D**) depict a progressive aggravation of *I. polycarpa* canker symptoms observed in the field.

### 2.2. Isolation of Pathogen and Determination of Pathogenicity

A total of 27 strains were isolated from the boundary between diseased and healthy tissues of *I. polycarpa*. Based on literature reviews, non-pathogenic strains were excluded. Preliminary identification, considering colony morphology and disease symptoms observed in the field, suggested that the pathogen belongs to the genus *Botryosphaeria*. Five representative strains (SQ1, SQ2, SQ3, SQ4, and SQ5) were selected for pathogenicity assays (Figure 2A). Branches inoculated with agar blocks showed no symptoms, whereas inoculation with these strains induced disease symptoms, including discoloration and depressed lesions at the inoculation sites (Figure 2B). Notably, branches inoculated with strain SQ5 exhibited significantly larger depressions compared to other strains, with some seedlings showing leaf wilting and abscission. Re-isolated colonies from infected tissues displayed morphological characteristics (color, shape, hyphal texture) consistent with the original strain, thus fulfilling Koch's postulates. Strain SQ5 was therefore identified as the most virulent pathogen.

### 2.3. Morphological and Cultural Characteristics

The SQ5 pathogen exhibits rapid growth, covering the entire plate within 5 to 6 days (Figure 4A,B). The colony is suborbicular with well-defined edges and dense mycelium. In the early stages of colony development, the mycelium appears white. By the third day, dark green or black pigmentation begins to emerge approximately 6.00 mm from the center of the colony. Simultaneously, aerial mycelium develops within the colony, characterized by thick, cotton-like growth that can stand erect. By the fifth day of incubation, the entire colony adopts a dark-green color. The pathogen was cultured in a spore-producing medium, leading to conidiophore production after approximately 14 days. Through light microscopy, the sporophore was observed to be colorless, rod-shaped, and producing conidia at its apex (Figure 4C). The conidia were spindle-shaped or fusiform, smooth, and non-septate (Figure 4D). Measurements of 30 conidia revealed that their size ranged from 23.678 to 30.332 μm in length and 5.132 to 6.795 μm in width, with an average size of 26.185 μm × 6.157 μm. These findings are consistent with the morphological characteristics of *B. dothidea* reported by Zheng et al. [17], He et al. [18], and Zhang et al. [19].

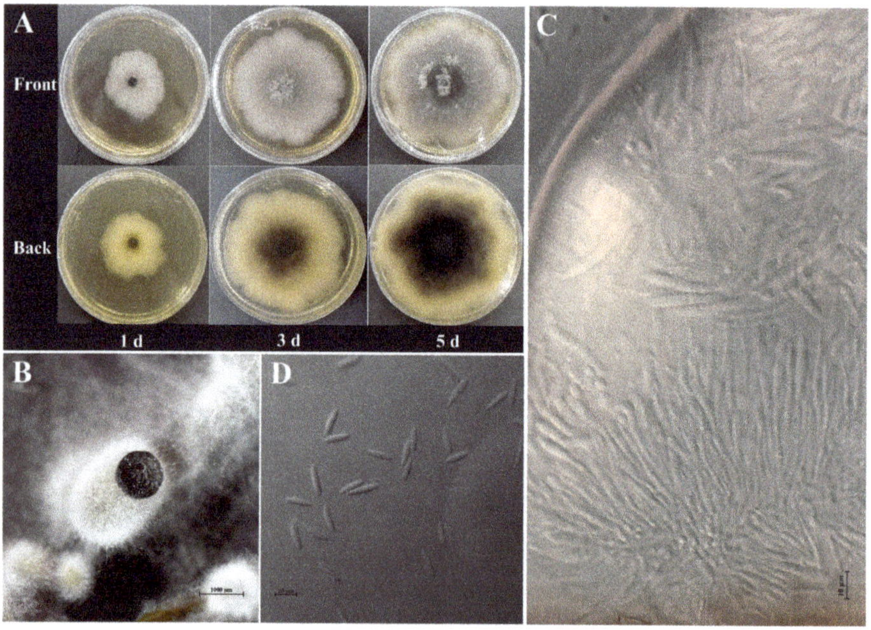


**Figure 4.** Morphological characteristics of *B. dothidea*. (**A**) Changes in colony morphology. (**B**) Section views of pycnidia. (**C**) Conidiophore. (**D**) Conidia.

*2.4. Molecular Identification and Phylogenetic Analyses*

To validate the morphological identification, we amplified ITS, β-*tubulin*, and *tef1-α* gene fragments from the SQ5 isolate. PCR amplification of the ITS region, β-*tubulin*, and *tef1-α* produced fragments of 523 bp, 432 bp, and 259 bp, respectively. The generated ITS, β-tubulin, and tef1-α sequences were submitted to GenBank (Table 1).

Preliminary alignment of these sequences with the NCBI database revealed that the SQ5 strain showed high similarity to *Botryosphaeria*. Phylogenetic analysis (Figure 5), based on the three-locus dataset (ITS + *tef1-α* + β-*tubulin*) and including 16 taxa, resulted in a dataset of 1326 characters. Of these, 156 characters were parsimony informative, 108 characters were parsimony uninformative, and 1062 characters were constant. A total of 27 trees were retained for the MP analysis (tree length = 313, CI = 0.939, RI = 0.972, RC = 0.913). Maximum Likelihood (ML) and Maximum Parsimony (MP) analyses based on the combined sequences revealed that SQ5 clustered with the previously reported *B. dothidea* JZG1 with 99% bootstrap support, clearly distinguishing it from other species of *Botryosphaeria*. Therefore, SQ5 was identified as *B. dothidea*.

**Table 1.** Sequences for phylogenetic analysis.

| Species | Isolate ID | *ITS* | *tef1-α* | *β-Tubulin* |
|---|---|---|---|---|
| **SQ5** | **STZ-1 = STZ3 = STZ2** | **OR523573** | **OR542956** | **OR542955** |
| *Neofusicoccum parvum* | CBS 123650 | KX464182 | KX464708 | KX464994 |
| *Neofusicoccum parvum* | CBS 145622 | MN611179 | MN623346 | MN623343 |
| *Neofusicoccum parvum* | CBS 145623 | MN611180 | MN623347 | MN623344 |
| *Botryosphaeria parva* * | CMW9077 | AY236939 | AY236884 | AY236913 |
| *Botryosphaeria parva* * | CMW9081 | AY236943 | AY236888 | AY236917 |
| *Botryosphaeria ribis* * | CMW7772 | AY236935 | AY236877 | AY236906 |
| *Botryosphaeria ribis* * | CMW7773 | AY236936 | AY236878 | AY236907 |
| *Botryosphaeria lutea* | KJ 93.52 = CMW992 = CMW992 | AF027745 | AY236894 | AY236923 |
| *Botryosphaeria lutea* | CMW9076 | AY236946 | AY236893 | AY236922 |
| *Botryosphaeria dothidea* * | CMW8000 | AY236949 | AY236898 | AY236927 |
| *Botryosphaeria dothidea* * | CMW7999 | AY236948 | AY236897 | AY236926 |
| *Botryosphaeria dothidea* | JZG1 | KU360149 | KU565872 | KU565871 |
| *Botryosphaeria dothidea* | JZB310203 | MN945382 | MT269886 | MT371073 |
| *Botryosphaeria dothidea* | JZB310200 | MN945379 | MT269883 | MT371070 |
| *Tiarosporella graminis* | CBS 118718 | KC769962 | KF531807 | KF531808 |

Sequences in this study are marked in bold. The ex-type strain is marked with *.

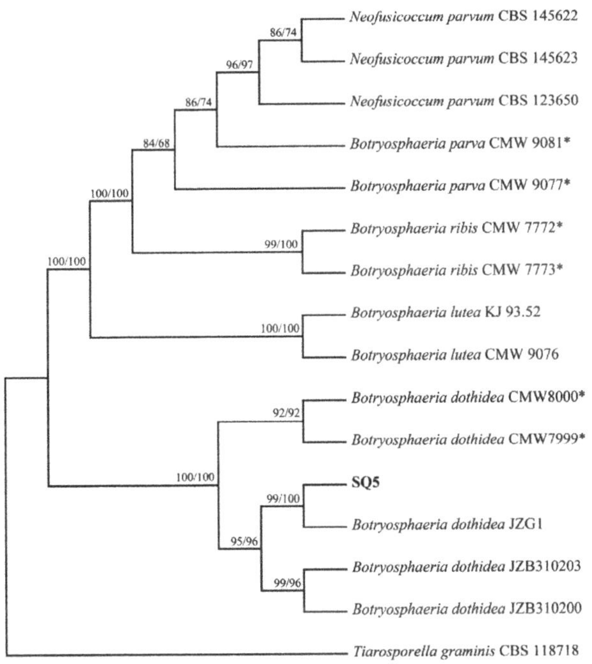


**Figure 5.** One of 27 equally parsimonious trees generated from maximum-parsimony analysis of the three-gene (ITS + tef1-α + β-tubulin) dataset. Numbers represent bootstrap values (MP/ML) which >70%. Isolates in bold were generated in this study. The ex-type strain is marked with *. *Tiarosporella graminis* was used to out-group.

### 2.5. Biological Characteristics

The growth of pathogens is closely linked to environmental conditions, prompting an investigation into their growth under various settings. The results indicated that the pathogen could thrive within a temperature range of 10–35 °C (Figure 6A), with significant differences observed ($p < 0.05$). After 5 days of incubation, the maximum colony diameter was recorded at 25 °C, reaching 90.00 mm, suggesting that 25 °C was the optimal growth temperature among the eight temperature gradients tested. However, growth inhibition occurred when temperatures fell below 10 °C or exceeded 35 °C. Temperatures of 40 °C and above significantly impaired pathogen growth (Figure 6B). No mycelial growth was observed after 10 min in a water bath at 60 °C and 65 °C, indicating a lethal temperature of approximately 60 °C for the pathogen. The pathogen demonstrated the ability to grow in media with pH values ranging from 3 to 11 (Figure 6C), with a notable difference ($p < 0.05$). Colonies in pH 3 media reached 90.00 mm by the third day, faster than in other treatment groups. Overall, pathogen colony size decreased as pH increased, indicating that the pathogen favored acidic conditions. Light exposure had a significant impact on pathogen growth (Figure 6D), with colonies exhibiting larger diameters in light than in darkness. Various carbon and nitrogen sources also significantly influenced pathogen growth, with the largest colony size observed when carboxymethylcellulose sodium was used as the carbon source (Figure 6E), and when casein tryptone served as the nitrogen source (Figure 6F).

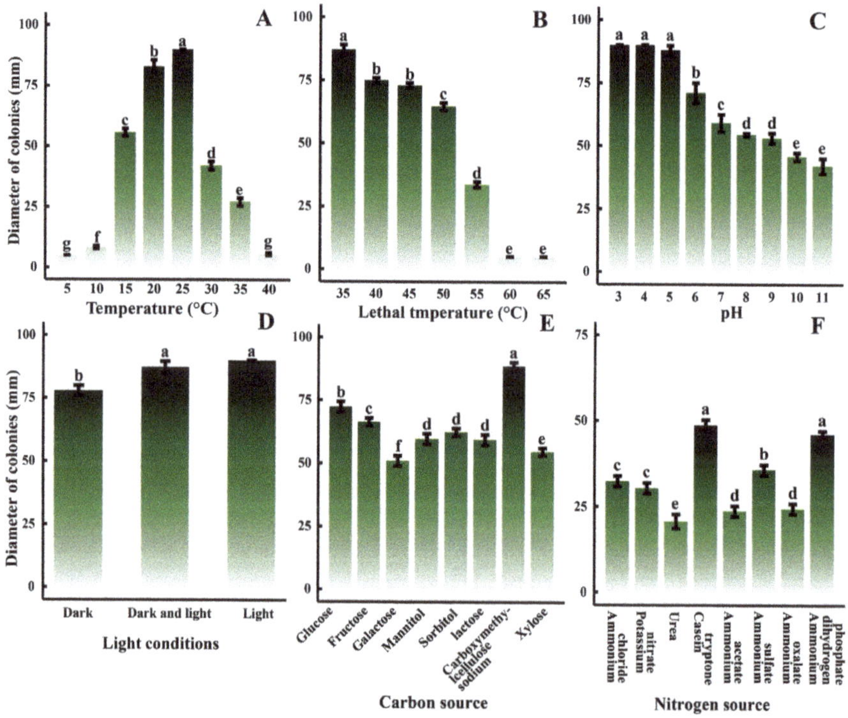


**Figure 6.** Effect of different environmental conditions on the growth of strain SQ5. (**A**) Temperature. (**B**) Lethal temperature. (**C**) pH; (**D**) Light conditions. (**E**) Carbon source. (**F**) Nitrogen source. Different lowercase letters represent significant differences ($p < 0.05$).

## *2.6. Screening of In Vitro Fungicides*

The concentration for 50% of the maximal effect (EC50) refers to the concentration required to achieve 50% of the maximum inhibitory effect. The inhibitory impact of each fungicide on pathogen growth can be quantitatively assessed by establishing a toxicity regression equation for each fungicide and calculating its EC50. The results (Table 2 and Figure S2) showed that 70% thiophanate-methyl exhibited the lowest EC50 at 0.0169 μg/mL, followed by 43% tebuconazole (EC50 = 0.0219 μg/mL), 20% xinjunan acetate (EC50 = 0.0271 μg/mL), and 40% difenoconazole (EC50 = 0.0954 μg/mL). These findings indicate that all four fungicides can exert a potent inhibitory effect on pathogen growth at relatively low concentrations.

**Table 2.** Toxicity test results of fungicides to pathogen.

| Number | Medicament | Toxicity Regression Equation | $R^2$ | EC50 (μg/mL) |
|---|---|---|---|---|
| 1 | 40% Thiabendazole | y = 0.9782x + 4.3588 | 0.9996 | 4.5237 |
| 2 | Azoxystrobin | y = 0.358x + 4.2849 | 0.7922 | 99.4288 |
| 3 | 30% Metalaxyl-M-metidaxyl | y = 2.1806x + 2.7381 | 0.9348 | 10.8964 |
| 4 | 43% Tebuconazole | y = 1.4269x + 7.369 | 0.9689 | 0.0219 |
| 5 | 20% Thiediazole copper | y = 1.9096x + 3.5989 | 0.9829 | 5.4164 |
| 6 | Lime sulfur | y = 0.2771x + 4.3538 | 0.5997 | 214.7880 |
| 7 | 20% Xinjunan acetate | y = 0.0316x + 6.0327 | 0.9376 | 0.0271 |
| 8 | 36%Kasugamycin-oxine-copper | y = 1.5122x + 4.0343 | 0.8071 | 4.3512 |
| 9 | 40% Difenoconazole | y = 0.8937x + 5.8912 | 0.9658 | 0.0954 |
| 10 | 6% kasugamycin | y = 1.8315x + 3.9717 | 0.9861 | 3.6433 |
| 11 | 70% Thiophanate-methyl | y = 0.8978x + 6.5902 | 0.9918 | 0.0169 |
| 12 | 47% Kasugamycin-copper oxychloride | y = 1.58x + 4.2467 | 0.9383 | 2.9976 |
| 13 | 27% Bromothalonil | y = 1.1286x + 4.0043 | 0.8658 | 7.6243 |
| 14 | 33.5% Oxine-copper | y = 0.3398x + 4.3141 | 0.9464 | 104.3615 |
| 15 | 25% Oligosaccharins-ethylicin | y = 0.9268x + 4.8564 | 0.9802 | 1.4286 |
| 16 | 30% Zincthiazole | y = 0.4042x + 4.534 | 0.9429 | 14.2200 |
| 17 | 80% Bordeaux mixture | y = 1.1139x + 4.9288 | 0.9561 | 1.1585 |

### 2.7. Validation of Field Efficacy

Field application results (Table 3) revealed that the morbidity of *I. polycarpa* treated with sprays containing 43% tebuconazole and 20% xinjunan acetate were 50.74% and 57.55%, respectively, both significantly lower than the control group (water). This suggests that these two fungicides have a strong inhibitory effect on the progression of canker lesions. The disease index provides a comprehensive measure of disease severity in plants. The results showed that the disease index for *I. polycarpa* sprayed with 43% tebuconazole was the lowest (11), while 70% thiophanate-methyl showed the highest disease index (16), both lower than the control group's index of 17. This indicates that the application of fungicides had a beneficial effect in alleviating the disease in *I. polycarpa*. When considering the overall control efficacy of the fungicides on *I. polycarpa* canker, the average efficacy values were as follows: 43% tebuconazole (35.29%) > 40% difenoconazole (25.53%) > 20% xinjunan acetate (11.76%) > 70% thiophanate-methyl (5.82%). No phytotoxic effects on the growth of *I. polycarpa* were observed during the trial.

**Table 3.** Control effect of *I. polycarpa* canker disease in the field.

| Medicament | Concentration (μg/mL) | Morbidity (%) | Disease Index | Average Efficacy (%) | Security |
|---|---|---|---|---|---|
| 43% Tebuconazole | 0.5 | 50.74 ± 5.23 c | 11 | 35.29 | Lower toxicity |
| 20% Xinjunan acetate | 2 | 57.55 ± 6.05 bc | 15 | 11.76 | Lower toxicity |
| 40% Difenoconazole | 4.5 | 70.00 ± 5.97 ab | 13 | 23.53 | Lower toxicity |
| 70% Thiophanate-methyl | 3.3 | 73.06 ± 1.06 a | 16 | 5.82 | Lower toxicity |
| Water (CK) | \ | 79.20 ± 1.49 a | 17 | \ | Innocuity |

Note: Different letters show significant differences among groups (mean ± SE, $p < 0.05$, n = 5).

## 3. Discussion

As a high-quality woody oilseed tree, *I. polycarpa* has been widely planted. This study investigated two plantation forests and revealed that canker disease poses a serious threat to *I. polycarpa*. It impairs growth in the early stages and leads to death in severe stages, potentially jeopardizing the large-scale cultivation of *I. polycarpa*. The major pathogen of *I. polycarpa* canker was identified through conventional tissue isolation and Koch's postulates. Morphological observations and multigene sequence analysis confirmed the pathogen as *Botryosphaeria dothidea*. *B. dothidea* is widespread and has been documented in several countries, including China [20], Japan [21], the United States [22], Iran [23], Spain [24], and Italy [25]. It infects a variety of woody host plants globally, such as *Cinnamomum camphora* [26], Taxus × media [27], *Persea americana* [3], *Ficus microcarpa* [28], *Carya cathayensis* [29], and *Prunus persica* [30]. Our observations also indicated that *I. polycarpa* canker primarily occurs on the tree's stem, and as the disease progresses, the cankers spread from the stem to the branches. This suggests that *I. polycarpa* canker is a systemic and highly damaging disease. Previous studies have shown that canker disease pathogens can induce tree dieback by causing functional failure of the phloem and cambium and carbon starvation in the xylem [31]. However, whether the mechanism by which *B. dothidea* kills *I. polycarpa* is similar remains to be further explored.

Based on field surveys, we observed a consistent pattern in the occurrence of *I. polycarpa* canker. The disease is primarily concentrated in months with favorable temperatures, sufficient light, and abundant rainfall, supporting the conclusion that the pathogen grows most rapidly at 25 °C and under full light conditions. *B. dothidea* is a weakly parasitic pathogen, characterized by latent infestation [32]. It typically resides in the soil as a saprophyte or within healthy plants as an endophyte. When favorable environmental conditions arise, it initiates plant infestation [33]. *Pistacia vera* is particularly susceptible to *B. dothidea* infection during the rainy season [34], and walnut shoot blight typically occurs in the warm, wet months of July and August [35]. Suitable environmental conditions promote pathogen growth and disease occurrence. Therefore, understanding the biological characteristics and growth patterns of pathogens is crucial for effective disease control and prevention in the field, and it provides a solid foundation for scientific management strategies.

In this study, the *B. dothidea* strain SQ5 exhibited strong aggressiveness toward *I. polycarpa*. Inoculation experiments showed rapid lesion expansion, tissue necrosis, and in some cases plant wilting and death, fulfilling Koch's postulates. Compared to reports of *B. dothidea* infecting other woody hosts, the SQ5 strain showed similarly high virulence, indicating its strong ability to colonize and damage *I. polycarpa* tissues. The aggressiveness may be attributed to its optimal growth at 25 °C, tolerance to acidic pH, broad carbon/nitrogen utilization, and a relatively high mycelial lethal temperature (60 °C), all of which enhance its adaptability and infection potential under field conditions. These traits make SQ5 a key pathogen of concern in *I. polycarpa* plantations and underscore the necessity for targeted control strategies.

In this study, we found that the *I. polycarpa* canker pathogen can utilize various carbon and nitrogen sources, indicating its capacity to thrive in diverse environmental conditions and survive under a wide range of circumstances, consistent with previous research [35–37]. However, we also observed some notable differences: the pathogen prefers an acidic environment but can still grow under alkaline conditions. This contrasts with the *B. dothidea* pathogen of *Osmanthus fragrans* leaf spot, which ceases growth at pH 3 [38], and the *B. dothidea* pathogen of *Photinia serrulata* leaf blight, which exhibits optimal growth at pH 6 [19]. Additionally, the mycelial lethal temperature in this study was 60 °C, higher than the mycelial lethal temperatures of 50 °C for the sweet cherry gum disease pathogen (*B. dothidea*) [39] and 52 °C for the passion fruit stem rot pathogen (*B. dothidea*) [40]. After plant infestation, during long-term parasitism and influenced by genetic, host, and environmental factors, the pathogenicity, cultural phenotype, and biological characteristics of the pathogen may exhibit significant variation between individuals [41–43]. This suggests that the *B. dothidea* strain in this study may have undergone mutation. Moreover, with global warming, the increasing frequency of warm weather may heighten the likelihood of *I. polycarpa* canker outbreaks.

When plants are infected with canker, they gradually exhibit a range of symptoms, including stem blight, ulcers, gum secretion, and fruit rot, as the pathogen's growth progresses, which adversely impacts plant health [44,45]. Currently, various measures are available to control the disease, including biological, chemical, and physical methods. Chemical control is widely used in agricultural production due to its simplicity and high efficiency [46]. Numerous fungicides are available on the market, each with varying levels of effectiveness against different plant diseases and pathogens. In vitro screening eliminates the influence of environmental factors and focuses on the fungicides' direct impact on pathogens, offering precise insights for pathogen control [47]. Li et al. [48] selected four fungicides from a pool of sixteen that showed significant inhibitory activity against the *Myrica rubra* twig blight pathogen. Similarly, Denman et al. [49] identified several fungicides, including tebuconazole, benomyl, and prochloraz, through laboratory experiments, providing valuable guidance for the field control of *Protea magnifica* canker. In this study, we screened four effective fungicides—70% thiophanate-methyl, 43% tebuconazole, 20% xinjunan acetate, and 40% difenoconazole—using a plate inhibition test. The EC50 values were 0.0169 μg/mL, 0.0219 μg/mL, 0.0271 μg/mL, and 0.0954 μg/mL, respectively. These fungicides also demonstrated strong inhibition of canker disease in other plants. For instance, Jiang et al. [50] found that 70% thiophanate-methyl had a strong inhibitory effect (EC50 = 9.3 μg/mL) on *B. dothidea* in their study on Chinese hickory canker. Tebuconazole inhibited the growth of *B. dothidea* (California pistachio blight) at low concentrations, with EC50 values ranging from 0.019 to 0.159 μg/mL [51]. Difenoconazole exhibited potent inhibitory activity against peach twig blight pathogens (*Phomopsis amygdali*, *Botryosphaeria obtusa*, and *Leucostoma persoonii*) with EC50 values of 0.289 μg/mL, 0.026 μg/mL, and 0.858 μg/mL, respectively [52]. The results from this study provide valuable insights for controlling *I. polycarpa* canker in the field.

During the in vitro fungicide screening, the fungicides showed strong inhibition through direct contact with the pathogens. However, the efficacy of these fungicides may be influenced by natural environmental conditions and the specific characteristics of the disease. Therefore, additional trials are required to assess and confirm their effectiveness against *I. polycarpa* canker under field conditions.

To assess the effectiveness of the fungicides under field conditions, we conducted a field validation trial. The results showed that, among the four fungicides tested, 43% tebuconazole and 20% Xinjunan acetate effectively reduced the incidence of *I. polycarpa* canker. Tebuconazole has been widely used to control plant cankers, reducing the occurrence of canker diseases caused by *Botryosphaeriales* fungi in apple and pear by 20% [53] and the incidence of Australian grapevine canker in the field by 41–65% [54]. In contrast, 40% difenoconazole had no significant effect on the incidence of *I. polycarpa* canker in our study, although previous research reported a 40% reduction in the incidence of apple necrotic rot [55]. In terms of tree health, the average efficacy of 43% tebuconazole and 40% difenoconazole against *I. polycarpa* canker in our study was 35.29% and 25.53%, respectively, which was lower than the control efficacy reported for other plant canker diseases. However, the field performance of 20% xinjunan acetate and 70% thiophanate-methyl was low, with control effects significantly lower than those observed in the laboratory tests. Olmo et al. [56] found that thiophanate-methyl significantly inhibited the apricot canker pathogen (*Botryosphaeriaceae* spp.) in vitro, with EC50 values ranging from 0.65 to 0.77 mg/L, and confirmed its effectiveness in reducing the length of apricot canker lesions and the pathogen isolation rate in field trials. The efficacy of fungicides is closely related to environmental factors such as temperature, moisture, and wind, which can reduce the concentration of active ingredients and affect treatment effectiveness [57]. Additionally, the timing and method of application can also influence efficacy. For example, spraying fungicides on peach trees in autumn can reduce canker incidence by 45–63%, while spring applications only reduce incidence by 10–28% [58]. Scraping lesions before applying fungicide has been shown to improve the control of *Litchi chinensis* canker [59]. Furthermore, Miller et al. [60] found that mixing fungicides can significantly improve efficacy. For instance, combining thiophanate-methyl with VitiSeal reduced peach canker pathogen viability more effectively than single applications. Our field trial, conducted in Zhengzhou City from July to August, occurred under climatic conditions characterized by high average daily temperatures and frequent rainfall. These factors may have compromised the effectiveness of the sprayed fungicides. Additionally, *I. polycarpa* canker originates from beneath the bark, where the pathogen resides, potentially reducing the fungicide's ability to penetrate and effectively control the disease. To date, no research has specifically addressed field control methods for *I. polycarpa* canker. Based on these findings, future research should focus on optimizing fungicide application methods and formulations to improve the field control of *I. polycarpa* canker.

## 4. Materials and Methods

### 4.1. Investigation of Field Diseases

Numerous *I. polycarpa* trees were planted at two locations: the Forest Seedling Breeding Engineering Technology Center of Henan Agricultural University (FSB) (113°38′39″, 34°48′2″) and the Science and Education Park of Henan Agricultural University (SEP) (113°36′, 34°50′24″) (Figure 6). The ages of the trees range from 3 to 15 years. Over time, as the duration of planting increased, symptoms of canker disease were observed at both locations, worsening each year. Therefore, between 2020 and 2021, we conducted field surveys to observe and record the timing and symptoms of *I. polycarpa* canker to better understand its incidence pattern.

### 4.2. Pathogen Collection and Isolation

During the *I. polycarpa* canker outbreak, 20 plants exhibiting typical canker symptoms were selected from each of the two sites for symptomatic sample collection. These samples were transported to the laboratory at the College of Forestry, Henan Agricultural University, for further analysis. For culture isolation, small sections (0.5 cm × 0.5 cm) of symptomatic stem tissue were surface-disinfected in 0.10% mercuric chloride for 30 s, transferred to 75% ethanol for another 30 s, rinsed four times in sterile water, dried on sterile absorbent paper, and then placed on potato dextrose agar (PDA) (Beijing Aoboxing Universeen Bio-tech Co., Ltd., Beijing, China) plates (90 mm in diameter) supplemented with 50 mg/L ampicillin (Sigma-Aldrich, St. Louis, MO, USA) to inhibit bacterial growth [2]. The plates were incubated at 25 ± 1 °C. The pathogen was purified by subculturing the colony edges after 4–6 days of growth, ensuring that the colonies had expanded but had not yet confluenced. The purified pathogen was then preserved using slant culture medium [61].

### 4.3. Pathogenicity Testing

To identify the primary pathogens causing *I. polycarpa* canker, the pathogenicity of the isolated strains was assessed following Koch's postulates. Pathogens from the slant medium were inoculated onto PDA plates and incubated at 25 ± 1 °C in the dark for 3 days to activate the culture. Subsequently, 5 mm mycelial plugs were extracted from the actively growing margin of the colony using a sterile hole punch. A small wound was made by removing a piece of bark (~5 mm in diameter) from the stem using a sterile scalpel. Healthy 3-year-old *I. polycarpa* seedlings were selected. An activated fungal plug was inserted into each wound. Each isolate was inoculated onto three seedlings, which were then wrapped tightly with sterile cotton saturated in sterile water and secured with parafilm. A 5 mm agar plug without the isolate served as the control. Symptoms were observed and recorded 21 days post-inoculation.

### 4.4. Morphological and Cultural Characterization

The pathogen was inoculated onto PDA plates and incubated at 25 ± 1 °C for 5 days, with colony morphology observed and recorded daily. The activated fungal isolate was also cultured on Murashige and Skoog (MS) culture medium, a plant tissue culture medium commonly used to promote sporulation in certain fungi [62]. Fresh cedar pine needles were rinsed in sterile water, evenly spread over the medium, and incubated in the dark at 25 ± 1 °C to promote sporophore and conidia formation [63]. The lengths and widths of 30 conidia were measured, and the shapes of both conidiophore and conidia were observed using an automated upright fluorescence microscope (Axio Imager M2, Carl Zeiss AG, Germany) with ZEN 3.8 software (Carl Zeiss AG, Oberkochen, Germany).

### 4.5. Molecular Identification

Approximately 100 mg of mycelium, activated at 25 ± 1 °C for 3 days, was collected and ground into a fine powder in a mortar using liquid nitrogen. Genomic DNA was then extracted using the Rapid Fungi Genomic DNA Isolation Kit (Sangon Biotech Co., Ltd., Shanghai, China), following the manufacturer's instructions. The ITS, β-*tubulin*, and translation elongation factor 1-α (*tef1-α*) genes were amplified and sequenced following established protocols [64–66]. PCR primers for the respective genes included: ITS1 (5′-TCCGTAGGTGAACCTGCGG-3′), ITS4 (5′-CCTCCGCTTATTGATATGC-3′) [67], Bt2a (5′-GGTAACCAAATCGGTGCTGCTTTC-3′), Bt2b (5′-ACCCTCAGTGTAGTGACCCTTGG C-3′) [68], EF1-728F (5′-CATCGAGAAGTTCGAGAAGG-3′), and EF1-986R (5′-ACTTGAAG GAACCCTTACC-3′) [69]. Although primer specificity assays were not conducted in this study, all primers used for the amplification of ITS, β-*tubulin*, and *tef1-α* regions were selected based on previously published and widely validated protocols. These primers have been extensively applied in the molecular identification of fungal pathogens, including *B. dothidea*.

PCR was conducted in a PCR apparatus (T100, Bio-Rad Laboratories, Hercules, CA, USA) with a 25 μL reaction mixture comprising 1 μL each of upstream and downstream primers, 2 μL fungal DNA template, 0.25 μL Taq DNA polymerase, 1 μL dNTPs, 2.5 μL 10X buffer, and 18.25 μL ddH$_2$O. The PCR amplification conditions were as follows: pre-denaturation at 94 °C for 10 min, denaturation at 94 °C for 30 s, annealing at 55–56 °C for 30 s, extension at 72 °C for 30 s, and a final extension at 72 °C for 10 min, repeated for 35 cycles. Agarose gel electrophoresis (1%) in 1 × TAE buffer was used to verify successful PCR amplification. Subsequently, DNA sequencing was performed by Beijing Liuhe Huada Gene Technology Co., Ltd., Beijing, China. Sequences obtained in this study were deposited in GenBank, and the accession numbers for the isolates are listed in Table 1.

### 4.6. Phylogenetic Analysis

ITS, β-*tubulin*, and *tef1-α* sequences of strains with high similarity to those in this study (including the ex-type strain) were selected based on correlation studies [17] and BLAST + 2.16.0 (Basic Local Alignment Search Tool) searches in the NCBI nucleotide database. DNA sequence datasets for *ITS*, β-*tubulin*, and *tef1-α* were aligned using MEGA 11 (Mega Limited, Auckland, New Zealand) [70], with manual adjustments as needed. Phylogenetic analyses of the combined ITS + *tef1-α* + β-*tubulin* dataset were conducted using PAUP v.4.0 (Sinauer Associates, Sunderland, MA, USA) [28]. Both Maximum Parsimony (MP) and Maximum Likelihood (ML) methods were employed. For the MP analysis, the heuristic search function and tree bisection and reconstruction (TBR) were used as the branch-swapping algorithm, with the branch-swapping option set to 'best tree' only.

Gaps were treated as 'missing', and characters were unordered and equally weighted [25], with Maxtrees limited to 100. Tree length (TL), consistency index (CI), retention index (RI), and rescaled consistency index (RC) were calculated. Clade support in both analyses was evaluated through 1000 bootstrap replicates. Additionally, *Tiarosporella graminis* (strain CBS 118718; GenBank ITS = KC769962, BT = KF531808, TEF = KF531807) was chosen as an outgroup [17].

### *4.7. Research on Biological Characteristics*

The pathogen was activated on PDA plates at 25 ± 1 °C for 3 days. Mycelial plugs (5 mm diameter) were excised from the actively growing colony margin using a sterile hole punch and subjected to the treatments described below. Each treatment included three replicates. Colony diameters were measured after 5 days of incubation using the cross-measurement method [25].

Effect of Temperature on Pathogen Growth: Mycelial plugs were cultured onto PDA plates and incubated in a constant temperature chamber at 5 °C, 10 °C, 15 °C, 20 °C, 25 °C, 30 °C, 35 °C, and 40 °C in the dark for 5 days.

Determination of the Lethal Temperature of the Pathogen: Each mycelial plug was placed in a 10 mL sterile centrifuge tube with 5.00 mL of sterile water. Tubes were immersed in a water bath at 40 °C, 45 °C, 50 °C, 55 °C, 60 °C, and 65 °C for 10 min [71]. After cooling to room temperature, each mycelial plug was inoculated onto a PDA plate and incubated at 25 ± 1 °C for 5 days.

Effect of pH on Pathogen Growth: The pH of the PDA medium was adjusted to 3, 4, 5, 6, 7, 8, 9, 10, and 11 using 1.00 mol/L HCl and 1.00 mol/L NaOH [72]. mycelial plugs were then inoculated onto PDA plates of each pH level and incubated at 25 ± 1 °C for 5 days.

Effect of Light on Pathogen Growth: Mycelial plugs were inoculated onto PDA plates and incubated at 25 ± 1 °C under four lighting conditions: complete light, complete darkness, and alternating 12-h cycles of light and darkness for 5 days.

Effect of Carbon and Nitrogen Sources on Pathogen Growth: Czapek-Dox medium was used as the base. Sucrose in the medium was replaced with glucose, fructose, galactose, mannitol, sorbitol, lactose, carboxymethylcellulose sodium, or xylose to create different carbon source media. Sodium nitrate was replaced with ammonium chloride, potassium nitrate, urea, casein tryptone, ammonium acetate, ammonium sulfate, ammonium oxalate, or ammonium dihydrogen phosphate to create different nitrogen source media [72]. Each mycelial plug was inoculated onto these carbon and nitrogen source media and incubated at 25 ± 1 °C for 5 days.

### *4.8. In Vitro Fungicide Screening*

To evaluate the effectiveness of fungicides in inhibiting pathogen growth, we tested 17 commercially available fungicides commonly used for controlling forest cankers (Table S1). A concentration gradient was established based on the recommended concentration for each fungicide, and each fungicide was mixed with PDA medium in a 1:9 ratio to create a fungicide-containing medium [28]. Activated mycelial plugs were then inoculated into each fungicide-containing medium, while PDA medium without fungicide served as the control. Three replicates were prepared for each treatment, and all plates were incubated in the dark at 25 ± 1 °C for 5 days. Colony diameters on each plate were measured using the crossover method, and the antifungal rate was calculated according to Equation (1).

$$IR = \frac{(DCC-DC)-(DTC-DC)}{DCC-DC} \times 100 \quad (1)$$

IR, Inhibition rate; DCC, Diameter of control colony; DC, Diameter of mycelial plug; DTC, Diameter of treated colonies.

### 4.9. Methods for Field Efficacy Verification

To assess the efficacy of fungicides screened in vitros, field trials were conducted. Diseased *I. polycarpa* trees of the same seed source, similar age, and growth were selected from the two trial sites, and the disease status of these sample trees was observed and recorded. Fungicides were chosen based on their low EC50 values obtained from in vitro toxicity measurements (Table 1), with concentrations set to achieve > 90% pathogen inhibition (Table S2). Five treatment groups were established according to the specifications in Table 4, with five replicates per treatment. Each group was kept separate to maintain experimental integrity.

**Table 4.** Test fungicide's information and concentration.

| Fungicide | Manufacturer | Concentration (mg/L) |
|---|---|---|
| 43% Tebuconazole | Bayer CropScience China Co., Ltd., Beijing, China | 0.50 |
| 20% Xinjunan acetate | Hebei Shenhua Pharmaceutical Co., Ltd., Shijiazhuang, China | 2.00 |
| 40% Difenoconazole | Shanghai Hulian Biopharmaceutical (Xiayi) Co., Ltd., Shanghai, China | 4.50 |
| 70% Thiophanate-methyl | Xi'an Dingsheng Biochemical Co., Ltd., Xi'an, China | 0.63 |
| Water | \ | \ |

The first application of the spray was carried out on 6 July 2022, using a hand-pressurized sprayer, followed by subsequent applications on 13, 20 and 27 July, with 1 L of each fungicide solution applied each time. The disease status of the sample trees was recorded on 3 August. Visual counts were conducted for spots on all sample trees, and the breast diameter of each tree was measured using a breast diameter caliper, along with the size of each spot. Disease ratings were calculated for each tree based on Table 5 [53,54,73]. Morbidity, disease index, and average efficacy were computed according to Equations (2), (3), and (4), respectively. Field efficacy tests were conducted following the EPPO guideline PP 1/135(4) for fungicide evaluation under practical conditions [74].

**Table 5.** Disease severity grading scale for field evaluation of *I. polycarpa* canker.

| Disease Rate | Representative Number | Grading Standard |
|---|---|---|
| I | 0 | No lesions |
| II | 1 | 1 to 10 spots on the main stem |
| III | 2 | The main stem has more than 10 spots and the width of the largest spot is less than 1/4 of the circumference of the tree. |
| IV | 3 | The main stem has more than 10 spots and the width of the largest spot is up to 1/4 but not more than 1/2 of the circumference of the tree. |
| V | 4 | The stem or branches are covered with disease scars, the maximum width of the spot is more than 1/2 the circumference of the tree, the branches and stems are crippled and the tree is weakened or even dead. |

$$M = \frac{NSBT - NSAT}{NSAT} \times 100 \quad (2)$$

M: Morbidity; NSBT: Number of spots before treatment; NSAT: Number of spots after treatment.

$$DI = \frac{\sum(NPPL \times LRV)}{TNPS \times HRV} \times 100 \quad (3)$$

DI: Disease index; NPPL: Number of plants per level; LRV: This level's representative value; TNPS: Total number of plants surveyed; HRV: Highest representative value.

$$AE = \frac{CGDI - TGDI}{CGDI} \times 100 \quad (4)$$

AE: Average efficacy; CGDI: Control group disease index; TGDI: Treatment group disease index.

*4.10. Statistical Analysis*

Data were processed and statistical analyses were performed using Excel 2021 (Microsoft Corp., Redmond, WA, USA). Normality and homogeneity of variances were assessed, and data were log2-transformed when necessary prior to analysis. Analysis of variance (ANOVA) was conducted using the 'aov' function in R (R Foundation for Statistical Computing, Vienna, Austria), followed by Tukey's honestly significant difference (HSD) test for post hoc comparisons. Visualizations were generated using the 'ggplot2' package in R. The spatial distribution of sampling sites was mapped using ArcGIS 10.7 (ESRI Inc., Redlands, CA, USA).

## 5. Conclusions

This study provides the first systematic report of canker disease in *I. polycarpa*, identifying *Botryosphaeria dothidea* as the causal agent based on morphological and molecular evidence and confirming its pathogenicity via Koch's postulates. The pathogen exhibits strong environmental adaptability and poses a significant threat to *I. polycarpa* plantations. In vitro and field evaluations revealed that 43% tebuconazole offers the most effective control among the tested fungicides. These findings lay a foundation for future research on disease epidemiology and management and highlight the need for integrated strategies to mitigate the impact of this emerging disease in expanding *I. polycarpa* cultivation.

**Supplementary Materials:** The following supporting information can be downloaded at: https://www.mdpi.com/article/10.3390/plants14091393/s1. Figure S1: Average monthly temperature and precipitation in Zhengzhou from 2021 to 2022; Figure S2: In vitro control using 17 different fungicides. The datasets generated and analyzed during the current study are available in the Figshare repository (available at: https://doi.org/10.6084/m9.figshare.28852358), Table S1: 17 Fungicide information. Table S2: Fungicide's antibactenial rate.

**Author Contributions:** Conceptualization, Q.C. and Z.L. (Zhi Li); methodology, J.F., Q.Y., X.C., L.F. and T.Z.; software, J.F., Q.Y., X.C., L.F. and T.Z.; validation, J.F., X.C., L.F. and T.Z.; formal analysis, J.F., Q.Y., X.C., L.F. and T.Z.; investigation, J.F., Q.Y., X.C., L.F. and T.Z.; resources, J.F., X.C., L.F., T.Z. and Z.L. (Zhi Li); data curation, J.F., X.C., L.F. and T.Z.; writing—original draft preparation, J.F.; writing—review and editing, Q.C., Z.L. (Zhi Li), Z.L. (Zhen Liu), Y.W. and X.G.; visualization, J.F.; supervision, Q.C. and Z.L. (Zhi Li); project administration, Q.C. and Z.L. (Zhi Li); funding acquisition, Q.C. and Z.L. (Zhi Li). All authors have read and agreed to the published version of the manuscript.

**Funding:** This research was funded by the Postgraduate Education Reform and Quality Improvement Project of Henan Province (YJS2024AL062/YJS2025AL63), the Henan Provincial Social Sciences Federation Research Project in China (SKL-2024-2179), the Henan Agricultural University's "Millions of Science and Education Service Action" Fund Project (2024SFBQW30), and the Henan Province Postdoctoral Research Project of China (202002053).

**Data Availability Statement:** All raw sequencing reads generated were deposited in the NCBI Sequence Read Archive (Nucleotide) under submission accession number OR523573, OR542956 and OR542955.

**Conflicts of Interest:** Authors Xuzhong Chen and Zhi Li were employed by the company Zhongyuan Environmental Protection Development Co., Ltd. The remaining authors declare that the research was conducted in the absence of any commercial or financial relationships that could be construed as a potential conflict of interest.

## References


1. López-López, Á.; Allende-Molar, R.; Correia, K.C.; Tovar-Pedraza, J.M.; Márquez-Zequera, I.; García-Estrada, R.S. First report of stem canker of Tomato caused by *Fusarium striatum* in Mexico. *Plant Dis.* **2021**, *105*, 497. [CrossRef]
2. Kovač, M.; Diminić, D.; Orlović, S.; Zlatković, M. *Botryosphaeria dothidea* and Neofusicoccum yunnanense causing canker and die-back of Sequoiadendron giganteum in Croatia. *Forests* **2021**, *12*, 695. [CrossRef]
3. Hernández, D.; García-Pérez, O.; Perera, S.; González-Carracedo, M.A.; Rodríguez-Pérez, A.; Siverio, F. Fungal pathogens associated with aerial symptoms of avocado (*Persea americana* Mill.) in Tenerife (Canary Islands, Spain) focused on species of the family Botryosphaeriaceae. *Microorganisms* **2023**, *11*, 585. [CrossRef] [PubMed]
4. Kenfaoui, J.; Lahlali, R.; Mennani, M.; Radouane, N.; Goura, K.; El Hamss, H.; El Ghadraoui, L.; Fontaine, F.; Tahiri, A.; Barka, E.A.; et al. Botryosphaeria dieback (Lasiodiplodia viticola): An imminent emerging threat to the Moroccan vineyards. *Plants* **2022**, *11*, 2167. [CrossRef] [PubMed]
5. Houston, D.R.; Valentine, H.T. Beech bark disease: The temporal pattern of cankering in aftermath forests of Maine. *Can. J. For. Res.* **1988**, *18*, 38–42. [CrossRef]
6. Ruess, R.W.; Winton, L.M.; Adams, G.C. Correction: Widespread mortality of trembling aspen (*Populus tremuloides*) throughout interior Alaskan boreal forests resulting from a novel canker disease. *PLoS ONE* **2021**, *16*, e0253996. [CrossRef]
7. Chen, W.Y.; Jiang, N.; Fan, X.L.; Guo, H.L.; Tian, C.M. Preliminary identification of pathogenic fungal groups from *Pinus armandii* canker in Gansu Province. *J. Northwest For. Univ.* **2020**, *35*, 147–152.
8. Lv, Q.; Zhang, Y.F.; Lin, R.Z.; Wang, H.M. Occurrence status of main forestry invasive species in China and their research trends. *Plant Prot.* **2022**, *48*, 21–38.
9. Wang, H.; Rana, S.; Li, Z.; Xu, G.; Wang, Y.; Cai, Q.; Li, S.; Sun, J.; Liu, Z. Morphological and anatomical changes during floral bud development of the trioecious *Idesia polycarpa* Maxim. *Braz. J. Bot.* **2022**, *45*, 679–688. [CrossRef]
10. Tian, X.X.; Fang, X.Z.; Du, M.H. Nutritional qualities and antioxidant properties of *Idesia polycarpa* Maxim. pulp oil and seed oil. *J. Chin. Cereals Oils Assoc.* **2020**, *35*, 91–95.
11. Xiang, X.; Wen, L.; Wang, Z.; Yang, G.; Mao, J.; An, X.; Kan, J. A comprehensive study on physicochemical properties, bioactive compounds, and emulsified lipid digestion characteristics of *Idesia polycarpa* var. vestita Diels fruits oil. *Food Chem.* **2023**, *404*, 134634. [CrossRef] [PubMed]
12. Zhou, D.; Zhou, X.; Shi, Q.; Pan, J.; Zhan, H.; Ge, F. High-pressure supercritical carbon dioxide extraction of *Idesia polycarpa* oil: Evaluation the influence of process parameters on the extraction yield and oil quality. *Ind. Crops Prod.* **2022**, *188*, 115586. [CrossRef]
13. Ai, Q.; Ai, J.; Wang, L. Oil grape (*Idesia polycarpa*) is included in the general food management by the state. *China For. Ind.* **2020**, *10*, 40–41.
14. Liu, Y. Analysis on the development and market prospect of oil grape (*Idesia polycarpa*) industry. *China For. Ind.* **2022**, *2*, 10–18.
15. Zhou, J.; Luo, J.X.; Yu, Y.H.; Liu, F.R.; Huang, S.H.; He, H.G. Major disease types of *Idesia polycarpa* in Sichuan Province and its prevention and control suggestions. *J. China West Norm. Univ.* **2023**, *44*, 237–243.
16. Zhu, Z.Y. *Propagation Technology and Development and Utilization of Idesia polycarpa*; Science Press: Beijing, China, 2010; pp. 206–222.
17. Zheng, X.R.; Zhang, M.J.; Shang, X.L.; Fang, S.Z.; Chen, F.M. Stem canker on Cyclocarya paliurus is caused by *Botryosphaeria dothidea*. *Plant Dis.* **2020**, *104*, 1032–1040. [CrossRef]
18. He, Y.; Qi, Z.; Li, S.; Zhu, H.; Hong, N.; Wang, G.; Wang, L. Molecular characterization of a new fusarivirus infecting *Botryosphaeria dothidea*, the causal agent of pear ring rot disease. *Arch. Virol.* **2022**, *167*, 1893–1897. [CrossRef]
19. Zhang, Q.; Zhang, Y.; Shi, H.; Huo, Y. *Botryosphaeria dothidea* causing leaf blight of *Photinia serrulata* in China. *Crop Prot.* **2023**, *174*, 106412. [CrossRef]
20. Dong, X.L.; Cheng, Z.Z.; Leng, W.F.; Li, B.H.; Xu, X.M.; Lian, S.; Wang, C.X. Progression of symptoms caused by *Botryosphaeria dothidea* on apple branches. *Phytopathology* **2021**, *111*, 1551–1559. [CrossRef]
21. Ono, Y.; Chatasiri, S.; Tanaka, E. Ochropsora staphyleae, a new rust pathogen of Japanese bladdernut, found in central Japan. *Mycoscience* **2020**, *61*, 62–64. [CrossRef]
22. Romero-Cuadrado, L.; López-Herrera, C.J.; Aguado, A.; Capote, N. Duplex real-time PCR assays for the simultaneous detection and quantification of Botryosphaeriaceae species Causing canker diseases in woody crops. *Plants* **2023**, *12*, 2205. [CrossRef] [PubMed]
23. Sohrabi, M.; Mohammadi, H.; León, M.; Armengol, J.; Banihashemi, Z. Fungal pathogens associated with branch and trunk cankers of nut crops in Iran. *Eur. J. Plant Pathol.* **2020**, *157*, 327–351. [CrossRef]
24. Gramaje, D.; Agustí-Brisach, C.; Pérez-Sierra, A.; Moralejo, E.; Olmo, D.; Mostert, L.; Damm, U.; Armengol, J. Fungal trunk pathogens associated with wood decay of almond trees on Mallorca (Spain). *Persoonia Mol. Phylogeny Evol. Fungi* **2012**, *28*, 1–13. [CrossRef] [PubMed]
25. Gusella, G.; Giambra, S.; Conigliaro, G.; Burruano, S.; Polizzi, G. Botryosphaeriaceae species causing canker and dieback of English walnut (Juglans regia) in Italy. *For. Pathol.* **2020**, *51*, e12661. [CrossRef]

26. Feng, X.; Lu, L.; Jia, Y.; Yan, X.; Zhongzhu, P.; Wang, G.; Qiu, C.; Wu, H. First report of trunk canker caused by *Botryosphaeria dothidea* on *Cinnamomum camphora* in China. *Plant Dis.* **2023**, *107*, 4030. [CrossRef]
27. Wang, Y.; Chen, J.; Lan, X.X.; Ren, M.; Zhang, B. First report of shoot and leaf blight caused by *Botryosphaeria dothidea* on *Taxus* × *media* in Sichuan Province, China. *Plant Dis.* **2023**, *107*, 2852. [CrossRef]
28. Fiorenza, A.; Aiello, D.; Costanzo, M.B.; Gusella, G.; Polizzi, G. A new disease for europe of *Ficus microcarpa* caused by Botryosphaeriaceae species. *Plants* **2022**, *11*, 727. [CrossRef]
29. Zhuang, C.J.; Wang, Q.W.; Wu, Q.Q.; Qiu, Z.L.; Xu, B.C.; Zhang, C.Q. Diversity of Botryosphaeriaceae species associated with Chinese hickory tree (*Carya cathayensis*) trunk cankers. *Plant Dis.* **2021**, *105*, 3869–3879. [CrossRef]
30. Tian, Y.; Zhao, Y.; Sun, T.; Wang, L.; Liu, J.; Ma, X.; Hu, B. Identification and characterization of *Phomopsis amygdali* and *Botryosphaeria dothidea* associated with peach shoot blight in Yangshan, China. *Plant Dis.* **2018**, *102*, 2511–2518. [CrossRef]
31. Xing, J.; Li, P.; Zhang, Y.; Li, J.; Liu, Y.; Lachenbruch, B.; Su, X.; Zhao, J. Fungal pathogens of canker disease trigger canopy dieback in poplar saplings by inducing functional failure of the phloem and cambium and carbon starvation in the xylem. *Physiol. Mol. Plant Pathol.* **2020**, *112*, 101523. [CrossRef]
32. Leal, C.; Trotel-Aziz, P.; Gramaje, D.; Armengol, J.; Fontaine, F. Exploring some factors conditioning the expression of Botryosphaeria dieback in grapevine for an integrated management of the disease. *Phytopathology* **2023**, *114*, 21–34. [CrossRef] [PubMed]
33. Slippers, B.; Wingfield, M.J. Botryosphaeriaceae as endophytes and latent pathogens of woody plants: Diversity, ecology and impact. *Fungal Biol. Rev.* **2007**, *21*, 90–106. [CrossRef]
34. Ntahimpera, N.; Driever, G.F.; Felts, D.; Morgan, D.P.; Michailides, T.J. Dynamics and pattern of latent infection caused by Botryosphaeria dothidea on Pistachio Buds. *Plant Dis.* **2002**, *86*, 282–287. [CrossRef] [PubMed]
35. López-Moral, A.; Lovera, M.; Raya, M.D.; Cortés-Cosano, N.; Arquero, O.; Trapero, A.; Agustí-Brisach, C. Etiology of branch dieback and shoot blight of English walnut caused by Botryosphaeriaceae and Diaporthe species in southern Spain. *Plant Dis.* **2020**, *104*, 533–550. [CrossRef]
36. Rezgui, A.; Ghnaya-Chakroun, A.B.; Vallance, J.; Bruez, E.; Hajlaoui, M.R.; Sadfi-Zouaoui, N.; Rey, P. Endophytic bacteria with antagonistic traits inhabit the wood tissues of grapevines from Tunisian vineyards. *Biol. Control* **2016**, *99*, 28–37. [CrossRef]
37. Chen, W.; Xie, F.; Tian, C.; Li, F.; Chou, G. Identification and characterization of Neocosmospora silvicola causing canker disease on *Pinus armandii* in China. *Plant Dis.* **2023**, *107*, 3026–3036. [CrossRef]
38. Li, Y.L.; Zhou, Z.; Yuan, C.Y.; Duan, P.L. Biological characteristics of the leaf spot lesions of Osmanthus fragrans and its resticide sensitivity. *North. Hortic.* **2014**, *19*, 117–120.
39. Zhang, Q.; Li, Y.; Yang, P.; Zhang, J.H.; Ma, Z.H.; Zhang, L.Z. Biological characteristics of Botryosphaeria dothoidea from sweet cherry. *Sci. Technol. Food Ind.* **2020**, *41*, 101–106.
40. Huang, Y.H.; Ning, P.; Huang, Y.G.; Ou, S.S.; Zhang, Y.X.; Tan, C.N.; Meng, C. Identification and biological characterization of stem rot pathogens from passion fruit. *Southwest China J. Agric. Sci.* **2022**, *35*, 105–112.
41. Sun, J.E.; Meng, C.-R.; Phillips, A.J.L.; Wang, Y. Two new Botryosphaeria (Botryosphaeriales, Botryosphaeriaceae) species in China. *MycoKeys* **2022**, *94*, 1–16. [CrossRef]
42. Li, T.; Li, N.; Lei, Z.; Zhang, C. Sensitivity and resistance risk of Botryosphaeria dothidea causing Chinese hickory trunk canker to fludioxonil. *Pestic. Biochem. Physiol.* **2023**, *194*, 105500. [CrossRef] [PubMed]
43. Manetti, G.; Brunetti, A.; Lumia, V.; Sciarroni, L.; Marangi, P.; Cristella, N.; Faggioli, F.; Reverberi, M.; Scortichini, M.; Pilotti, M. Identification and characterization of Neofusicoccum stellenboschiana in branch and twig dieback-affected *Olea europaea* trees in Italy and comparative pathogenicity with *N. mediterraneum*. *J. Fungi* **2023**, *9*, 292. [CrossRef] [PubMed]
44. Geng, X.; Liu, Y.; Li, J.; Li, Z.; Shu, J.; Wu, G. Identification and characterization of Nectria pseudotrichia associated with camellia canker disease in China. *Forests* **2022**, *13*, 29. [CrossRef]
45. Dobry, E.; Rutter, M.A.; Campbell, M.A. The fungal pathogen Gnomoniopsis castaneae induces damaging cankers in multiple domestic Fagaceae Species. *Phytopathology* **2023**, *113*, 1817–1821. [CrossRef]
46. Feng, H.; Wang, C.; Yang, H.; Tang, L.; Han, P.-L.; Liang, J.; Huang, L. Apple valsa canker: Insights into pathogenesis and disease control. *Phytopathol. Res.* **2023**, *5*, 45. [CrossRef]
47. Tudela, M.A.A.; Lutz, M.C.; Giménez, G.N.; Del Brío, D.; Di Masi, S.N.; Pose, G.N.; Molina, J.P.E. Efficacy of fungicides against brown spot of pear in Argentina. *Crop Prot.* **2023**, *174*, 106425. [CrossRef]
48. Li, W.; Hu, M.; Xue, Y.; Li, Z.; Zhang, Y.; Zheng, D.; Lu, G.; Wang, J.; Zhou, J. Five fungal pathogens are responsible for bayberry twig blight and fungicides were screened for disease control. *Microorganisms* **2020**, *8*, 689. [CrossRef]
49. Denman, S.; Crous, P.W.; Sadie, A.; Wingfield, M.J. Evaluation of fungicides for the control of Botryosphaeria protearum on *Protea magnifica* in the Western Cape Province of South Africa. *Australas. Plant Pathol.* **2004**, *33*, 97–102. [CrossRef]
50. Jiang, H.; Rao, Y.; Li, M.; Wang, Y. Antifungal activity of rapamycin on Botryosphaeria dothidea and its effect against Chinese hickory canker. *Pest Manag. Sci.* **2020**, *77*, 425–431. [CrossRef]

51. Ma, Z.; Morgan, D.P.; Felts, D.; Michailides, T.J. Sensitivity of Botryosphaeria dothidea from California pistachio to tebuconazole. *Crop Prot.* **2002**, *21*, 829–835. [CrossRef]
52. Froelich, M.H.; Schnabel, G. Investigation of fungi causing twig blight diseases on peach trees in South Carolina. *Plant Dis.* **2019**, *103*, 705–710. [CrossRef]
53. Sosa, M.C.; Lutz, M.C.; Lódolo, X.V.; Basso, C.N. In vitro and in vivo activity of chemical fungicides and a biofungicide for the control of wood diseases caused by Botryosphaeriales fungi in apple and pear. *Int. J. Pest Manag.* **2022**, *68*, 328–338. [CrossRef]
54. Pitt, W.M.; Sosnowski, M.R.; Huang, R.; Qiu, Y.; Steel, C.C.; Savocchia, S. Evaluation of fungicides for the management of Botryosphaeria canker of grapevines. *Plant Dis.* **2012**, *96*, 1303–1308. [CrossRef] [PubMed]
55. Song, Y.; Li, L.; Li, C.; Lu, Z.; Men, X.; Chen, F. Evaluating the sensitivity and efficacy of fungicides with different modes of action against Botryosphaeria dothidea. *Plant Dis.* **2018**, *102*, 1785–1793. [CrossRef] [PubMed]
56. Olmo, D.; Gramaje, D.; Armengol, J. Evaluation of fungicides to protect pruning wounds from Botryosphaeriaceae species infections on almond trees. *Phytopathol. Mediterr.* **2017**, *56*, 77–86.
57. Hou, Y.; Guo, Y.; Wang, L.; He, S.; Zheng, W.; Liu, S.; Xu, J. Impact of phenamacril on the growth and development of Fusarium pseudograminearum and control of crown rot of wheat. *Plant Dis.* **2023**, *107*, 3843–3850. [CrossRef]
58. Lalancette, N.; Robison, D.M. Effect of fungicides, application timing, and canker removal on incidence and severity of constriction canker of peach. *Plant Dis.* **2002**, *86*, 721–728. [CrossRef]
59. Liao, M.J.; Zhang, Z.H.; Xi, P.G.; Wu, Y.Z.; Zhang, H.L.; Jiang, Z.D. Incidence and control of litchi stem rot in Zengcheng. *China Trop. Agric.* **2021**, *4*, 30–34.69.
60. Miller, S.T.; Sterle, D.; Minas, I.S.; Stewart, J.E. Exploring fungicides and sealants for management of Cytospora plurivora infections in western Colorado peach production systems. *Crop Prot.* **2021**, *146*, 105654. [CrossRef]
61. Zhang, C.Q.; Xu, B.C. First report of canker on Chinese hickory (*Carya cathayensis*) caused by Botryosphaeria dothidea in China. *Plant Dis.* **2011**, *95*, 1319. [CrossRef]
62. Murashige, T.; Skoog, F. A revised medium for rapid growth and bio assays with tobacco tissue cultures. *Physiol. Plant.* **1962**, *15*, 473–497. [CrossRef]
63. Liu, J.; Zhang, L.Y.; Wang, H.Y.; Liu, N.; Lian, S.; Xu, X.M.; Li, B.H. The effect of temperature and moisture on colonization of apple fruit and branches by Botryosphaeria dothidea. *Phytopathology* **2022**, *112*, 1698–1709. [CrossRef] [PubMed]
64. White, T.J.; Bruns, T.; Lee, S.; Taylor, J. Amplification and direct sequencing of fungal ribosomal RNA Genes for phylogenetics. *PCR Protoc.* **1990**, *18*, 315–322.
65. Glass, N.L.; Donaldson, G.C. Development of primer sets designed for use with the PCR to amplify conserved genes from filamentous ascomycetes. *Appl. Environ. Microbiol.* **1995**, *61*, 1323–1330. [CrossRef] [PubMed]
66. Alves, A.; Crous, P.; Correia, A.; Phillips, A. Morphological and molecular data reveal cryptic speciation in Lasiodiplodia theobromae. *Fungal Divers.* **2008**, *28*, 1–13.
67. Yu, L.; Wang, L.F.; Zhao, J.R.; Xu, S.G.; Gao, D.; Zheng, J.F. First report of *Botryosphaeria dothidea* causing canker and dieback disease of Helwingia chinensis in China. *Plant Dis.* **2012**, *96*, 1821. [CrossRef]
68. Al-Naemi, F.A.; Nishad, R.; Ahmed, T.A.; Radwan, O. First report of *Thielaviopsis punctulata* causing black scorch disease on date palm in qatar. *Plant Dis.* **2014**, *98*, 1437. [CrossRef]
69. Tang, W.; Ding, Z.; Zhou, Z.Q.; Wang, Y.Z.; Guo, L.Y. Phylogenetic and pathogenic analyses show that the causal agent of apple ring rot in China is *Botryosphaeria dothidea*. *Plant Dis.* **2012**, *96*, 486–496. [CrossRef]
70. Tamura, K.; Peterson, D.; Peterson, N.; Stecher, G.; Nei, M.; Kumar, S. MEGA5: Molecular evolutionary genetics analysis using maximum likelihood, evolutionary distance, and maximum parsimony methods. *Mol. Biol. Evol.* **2011**, *28*, 2731–2739. [CrossRef]
71. Shi, J.; Sang, W.J.; Lu, Y.H.; Zhang, D.P.; Kong, F.; Shou, A.F.; Yang, M.M. Identification of tobacco anthracnose pathogen and its biological characteristics in Guangxi. *Chin. Tob. Sci.* **2023**, *44*, 70–78.
72. Cui, L.; Yang, C.; Jin, M.; Wei, L.; Yang, L.; Zhou, J. Identification and biological characterization of a new pathogen that causes potato scab in Gansu Province, China. *Microb. Pathog.* **2021**, *161*, 105276. [CrossRef] [PubMed]
73. Shi, D.Y.; Wei, B.F.; Li, H.P.; Zeng, J.Y.; Wang, H.L.; Li, Y.L. Evaluation on the control effect of four biological fungicides against verticillium wilt of Cotinus. *For. Pest Dis.* **2023**, *42*, 46–50.
74. European and Mediterranean Plant Protection Organization (EPPO). EPPO Standard PP 1/135(4): Efficacy Evaluation of Plant Protection Products—Phytotoxicity Assessment. *EPPO Bull.* **2014**, *44*, 265–273.

*plants*

MDPI

*Article*

# Effective Bud Induction of *Acacia mangium* and *A. auriculiformis* Without $KNO_3$ and $NH_4NO_3$ in Media

**Lin Sun, Yanping Lu and Liejian Huang ***

Research Institute of Tropical Forestry, Chinese Academy of Forestry, Guangzhou 510520, China; sunlin721@163.com (L.S.); 15067100402@163.com (Y.L.)

* Correspondence: 13802987948@163.com; Tel.: +86-13802987948

**Abstract**

Stem segments of *Acacia mangium* and *A. auriculiformis* containing full axillary buds were used to study the effects of reduced amounts of the main nitrogen source in the growth media. This condition, referred to as nitrogen deficiency in this article and denoted as -N, involved the omission of ammonium nitrate and potassium nitrate from MS media, and its impact on bud induction was assessed. The results show that in media lacking nitrogen, the bud induction rate, contamination rate, browning rate, stem length, and leaf number of induced buds of *A. mangium* and *A. auriculiformis* varied depending on the different culture media used. The optimal bud induction medium for *A. mangium* and *A. auriculiformis* was as follows: 1/4MS (-N) + 1.0 mg·L$^{-1}$ 6-BA + 0.2 g·L$^{-1}$ chlorothalonil + 5 g·L$^{-1}$ AGAR. The bud induction rates were 72.6% and 100.0%, respectively. There were no significant differences in the rooting rates of the induced buds between the -N treatment and the complete nutrient treatment. We found that the buds induced in the -N media did not show obvious symptoms of nitrogen deficiency, and their growth status was not significantly different from those induced in the complete nutrient media, which indicates that nitrogen is not essential for the bud induction of *A. mangium* and *A. auriculiformis*. The results of this study provide an important reference for conducting related research on other plants and have are greatly significant for the sustainable development of tissue culture technology in the future.

**Keywords:** *Acacia*; tissue culture; nitrogen deficiency; bud induction; rooting; 6-BA

## 1. Introduction

*Acacia mangium* and *A. auriculiformis* were introduced in China in the 1960s [1,2]. Due to their economic benefits and ecological value (such as good adaptability, rapid growth, high yield, developed nodules, nitrogen fixation ability, long fibers, etc.), *A. mangium* and *A. auriculiformis* became important tree species for barren mountain afforestation, landscaping, and fuelwood in the south of China [3–6].

*Acacia* propagates through both sexual and asexual methods. Its seed progeny exhibit significant genetic differentiation in traits such as tree shape, resistance, wood properties, and other aspects. Cutting propagation of *Acacia* is greatly affected by the seasons and other factors. For example, excessively high temperatures in summer can enhance the respiratory function of cuttings, consume excessive nutrients, and also easily lead to bacterial infection at the cut ends, which is not conducive to root growth. At the same time, the rate of water evaporation is high, and the cuttings are prone to water loss and withering. As a result, to a certain extent, the plantation and utilization of *Acacia* species are limited [7,8]. Therefore, it is very important to establish efficient propagation technology to provide a large number of high-quality seedlings and meet the demands of *Acacia* plantation and utilization.

Tissue culture is an essential asexual propagation technology that has the advantages of controllable, convenient management, high reproductive efficiency, and the ability to maintain excellent parental traits [9]. In recent years, researchers have performed extensive research on *Acacia* tissue culture [10–16], which has supported the plantation and utilization of *Acacia*.

Using a suitable medium is the key to successfully establishing a tissue culture technique. Widely used media include the MS and improved MS media, which consist of a large number of elements, trace elements, iron salts, organic matter, and more. Each component plays an important role in the growth and development of plants. Among them, nitrogen is indispensable, serving as the basic component of chlorophyll, nucleic acid, protein, and more. It regulates many physiological processes in plants through signal transduction pathways, ultimately regulating plant growth and development [17–19]. During tissue culture, the exogenous nitrogen supplied in the medium comes from nitrate nitrogen (potassium nitrate) and ammonia nitrogen (ammonium nitrate) [20–22]. As potassium nitrate and ammonium nitrate are the main raw materials for the production of explosives, their purchase and use have been strictly controlled in most countries and regions in recent years, with China has also implementing strict supervision. They are classified as explosive, dangerous goods in the Identification of Major Hazard Sources of Hazardous Chemicals [23], which seriously impact their promotion and application in the tissue culture industry. Furthermore, few studies have examined the influence of nitrogen on *Acacia* tissue culture seedlings and the response to nitrogen in the process of *Acacia* tissue culture.

Therefore, in order to solve the above-mentioned problems, this study used *A. mangium* and *A. auriculiformis* as experimental materials to study the effect of nitrogen on bud induction. Through adjusting the proportion of MS medium and the concentration of exogenous cytokinin 6-BA, the optimal bud induction medium for *A. mangium* and *A. auriculiformis* under nitrogen-deficiency conditions was identified. In addition, the induced buds were subjected to a rooting culture to compare the differences in rooting between *A. mangium* and *A. auriculiformis* under nitrogen-deficient conditions and complete nutrient conditions. This evaluation of the effects of nitrogen-deficient bud induction lays the foundation for relevant research of *Acacia* nitrogen-deficiency culture. In addition, the research results will be very helpful for the sustainable development of tissue culture technology in the future.

## 2. Results

### 2.1. *Effects of N and 6-BA on Bud Induction*

Under the -N treatment, the bud induction rate and browning rate of *A. mangium* were not significantly different between the different media used. The bud induction rate decreased with the increase in the concentration of the basic medium. The bud induction rates, from highest to lowest, were 1/8MS = 1/4MS > 1/2MS > MS, with the highest induction rate reaching 66.9%. The browning rates in all media used were lower than 20%, and the contamination rates were lower than 5%. The contamination rate in the 1/4MS medium was significantly higher than that in the 1/2MS medium (Figure 1).

Under the F treatment, with the increase in the concentration of the basic medium, the bud induction rates of *A. mangium* decreased significantly, and the browning rates increased significantly. The bud induction rates, from highest to lowest, were 1/8MS > 1/4MS > 1/2MS > MS, with the highest induction rate reaching 88.9%. The browning rates, from highest to lowest, were MS > 1/2MS > 1/4MS > 1/8MS, with the highest browning rate reaching 78.5%. The contamination rates in all media were lower than 10%, and the contamination rate in the 1/2MS medium was significantly higher than that in the 1/4MS medium and 1/8MS medium (Figure 1).

Under the -N treatment, with the increase in the concentration of the basic medium, the bud induction rates of *A. auriculiformis* increased first and then decreased. The bud induction rate in the 1/4MS medium was the highest, at 93.8%, and the induction rate in the MS medium was significantly lower than that in the other treatments. There were no significant differences in the browning rate or contamination rate of the different media (Figure 2).

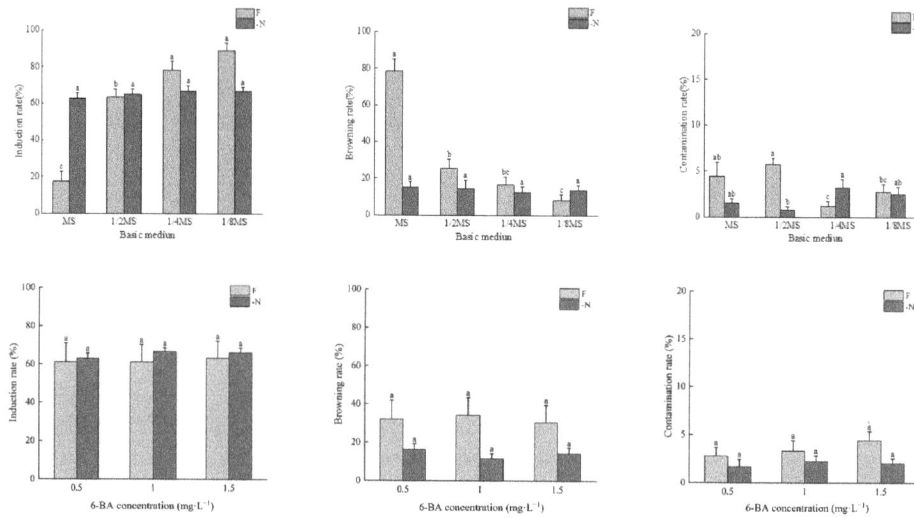


**Figure 1.** The effects of N and 6-BA on bud induction of *A. mangium*. Different lower-case letters indicate significant differences at $p < 0.05$.

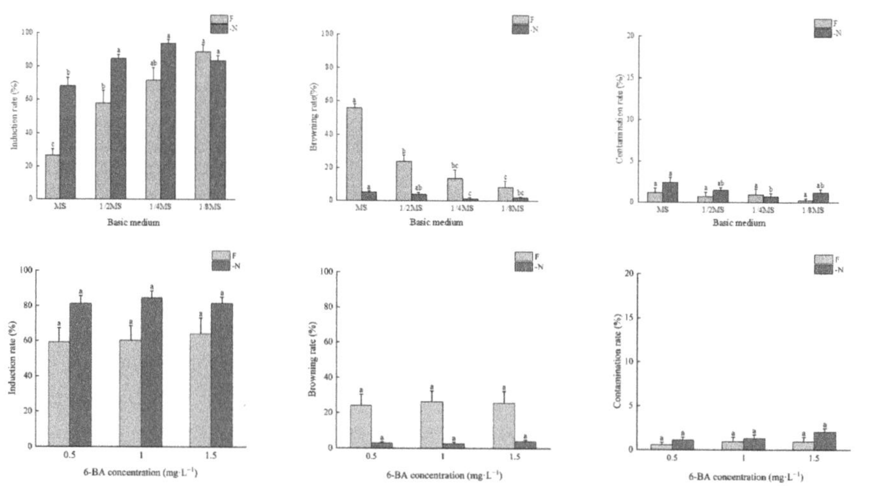


**Figure 2.** The effects of N and 6-BA on bud induction of *A. auriculiformis*. Different lower-case letters indicate significant differences at $p < 0.05$.

Under the F treatment, with the increase in the concentration of the basic medium, the bud induction rates of *A. auriculiformis* decreased significantly, and the browning rates increased significantly. The bud induction rates, from highest to lowest, were 1/8MS > 1/4MS > 1/2MS > MS, with the highest induction rate reaching 88.6%. The browning rates, from highest to lowest, were MS > 1/2MS > 1/4MS >1/8MS, with the highest browning rate reaching 55.6%. There were no significant differences in the contamination rates in the different media, which were all lower than 5% (Figure 2).

Under the different concentrations of 6-BA, there were no significant differences in the bud induction rates, browning rates, or contamination rates of *A. mangium* and *A. auriculiformis* between the nitrogen-deficient media and complete nutrient media. Under the same 6-BA treatment concentration, the bud induction rates of *A. mangium* and the bud induction rates and contamination rates of *A. auriculiformis* were all higher in the nitrogen-deficient media than in the complete nutrient media. Conversely, the browning rates and contamination rates of *A. mangium* and the browning rates of *A. auriculiformis* were all higher in the complete nutrient media than in the nitrogen-deficient media (Figures 1 and 2).

### 2.2. Effects of N on Stem Length and Leaf Number of Induced Buds

The results of multiple comparative analysis show that the lack of nitrogen in the media had a certain effect on the bud induction of *A. mangium* and *A. auriculiformis*. Among the 12 -N treatments, *A. mangium* and *A. auriculiformis* had the highest bud induction rates with treatment 8- (-N), with 72.6% and 100.0%, respectively. Among the 12 F treatments, the bud induction rate of *A. mangium* was the highest in treatment 10- (F), with 91.9%; the bud induction rate of *A. auriculiformis* was the highest in treatment 12- (F), with 94.1%. The highest bud induction rate of *A. auriculiformis* treated with -N was higher than that of the F treatment. The stem lengths of the induced *A. mangium* and *A. auriculiformis* buds in the complete nutrient media were higher than those in the nitrogen-deficient media (Table 1), indicating that suitable media can normally induce germination of *Acacia* macrophylla explants under nitrogen-deficiency conditions, but it will affect the elongation and growth of the buds to some extent. After 40 days, it was observed that the number of leaves on buds induced in the nitrogen-deficient culture was lower than those induced in the complete nutrient culture, and the leaf color was light green; however, no symptoms of nitrogen deficiency were observed (Figure 3). These results indicate that the bud induction rate and bud growth of *A. mangium* and *A. auriculiformis* were not significantly negatively affected under the absence of nitrogen. Comprehensive analysis showed that the optimal nitrogen-deficient medium for both *A. mangium* and *A. auriculiformis* bud induction was 1/4MS (-N) + 1.0 mg·L$^{-1}$ 6-BA.

**Table 1.** The stem lengths and leaf numbers of induced *A. mangium* and *A. auriculiformis* buds.

| No. | Treatments | *A. mangium* | | | *A. auriculiformis* | | |
|---|---|---|---|---|---|---|---|
| | | Bud Induction Rate (%) | Stem Length (mm) | Leaf Number | Bud Induction Rate (%) | Stem Length (mm) | Leaf Number |
| 1 | MS (-N), 0.5 mg/L 6-BA | 65.2 b | 7.0 b | 2.5 b | 64.4 b | 7.2 b | 2.2 b |
| | MS (F), 0.5 mg/L 6-BA | 12.6 c | 7.1 b | 2.5 b | 29.6 c | 8.9 b | 2.1 b |
| 2 | MS (-N), 1.0 mg/L 6-BA | 66.7 b | 8.2 b | 2.9 b | 70.4 b | 8.3 b | 2.7 b |
| | MS (F), 1.0 mg/L 6-BA | 17.1 c | 8.8 b | 2.6 b | 27.4 c | 11.2 ab | 2.5 b |
| 3 | MS (-N), 1.5 mg/L 6-BA | 56.3 b | 8.1 b | 3.1 ab | 69.6 b | 8.4 b | 2.8 b |
| | MS (F), 1.5 mg/L 6-BA | 22.2 c | 8.5 b | 3.0 b | 23.0 c | 11.1 ab | 2.8 b |
| 4 | 1/2MS (-N), 0.5 mg/L 6-BA | 64.5 b | 8.2 b | 3.0 ab | 83.0 ab | 8.2 b | 2.2 b |
| | 1/2MS (F), 0.5 mg/L 6-BA | 61.5 b | 11.5 ab | 3.7 ab | 55.5 bc | 12.4 ab | 2.9 b |
| 5 | 1/2MS (-N), 1.0 mg/L 6-BA | 60.7 b | 8.5 b | 3.2 ab | 83.0 ab | 9.3 b | 2.9 b |
| | 1/2MS (F), 1.0 mg/L 6-BA | 60.7 b | 15.0 a | 4.7 a | 57.1 b | 14.6 a | 3.4 ab |
| 6 | 1/2MS (-N), 1.5 mg/L 6-BA | 68.9 b | 9.2 b | 3.6 ab | 87.4 ab | 9.8 b | 2.6 b |
| | 1/2MS (F), 1.5 mg/L 6-BA | 67.4 b | 14.4 ab | 4.6 ab | 60.7 b | 14.7 a | 3.7 ab |
| 7 | 1/4MS (-N), 0.5 mg/L 6-BA | 60.0 b | 7.4 b | 2.8 b | 95.5 ab | 8.9 b | 2.4 b |
| | 1/4MS (F), 0.5 mg/L 6-BA | 79.3 ab | 14.1 ab | 4.5 ab | 66.7 b | 13.4 ab | 3.1 ab |
| 8 | 1/4MS (-N), 1.0 mg/L 6-BA | 72.6 ab | 8.5 b | 3.1 ab | 100.0 a | 9.8 b | 2.8 b |
| | 1/4MS (F), 1.0 mg/L 6-BA | 77.8 ab | 14.5 ab | 4.4 ab | 69.6 b | 13.8 ab | 3.4 ab |
| 9 | 1/4MS (-N), 1.5 mg/L 6-BA | 68.1 b | 7.9 b | 3.4 ab | 85.9 ab | 10.5 ab | 3.0 ab |
| | 1/4MS (F), 1.5 mg/L 6-BA | 77.8 ab | 14.1 ab | 4.4 ab | 78.5 ab | 14.0 ab | 3.8 a |
| 10 | 1/8MS (-N), 0.5 mg/L 6-BA | 63.0 b | 7.8 b | 2.9 b | 82.2 ab | 9.5 b | 2.6 b |
| | 1/8MS (F), 0.5 mg/L 6-BA | 91.9 a | 13.4 ab | 4.3 ab | 85.2 ab | 13.6 ab | 3.2 ab |
| 11 | 1/8MS (-N), 1.0 mg/L 6-BA | 66.6 b | 7.8 b | 3.2 ab | 85.2 ab | 9.4 b | 2.5 b |
| | 1/8MS (F), 1.0 mg/L 6-BA | 88.9 ab | 13.8 ab | 4.3 ab | 86.7 ab | 14.7 a | 3.4 ab |
| 12 | 1/8MS (-N), 1.5 mg/L 6-BA | 71.1 ab | 9.0 b | 3.5 ab | 83.0 ab | 10.3 ab | 2.8 b |
| | 1/8MS (F), 1.5 mg/L 6-BA | 85.9 ab | 14.6 ab | 4.3 ab | 94.1 ab | 14.8 a | 3.7 ab |

Different lower-case letters indicate significant differences at $p < 0.05$. Data are means ± standard error.

(a) (b) (c) (d)

**Figure 3.** *A. mangium*- and *A. auriculiformis*-induced buds. (**a**) *A. mangium* buds induced in treatment 8- (-N). (**b**) *A. mangium* buds induced in treatment 10- (F). (**c**) *A. auriculiformis* buds induced in treatment 8- (-N). (**d**) *A. auriculiformis* buds induced in treatment 12- (F). Scale bars (top right-hand corner in each picture) = 1 cm.

### 2.3. Rooting of Buds Induced in -N and F Treatments

*A. mangium* and *A. auriculiformis* buds induced under the -N and F treatments were subsequently cultured for rooting. White root primordia began to appear at the bottom of the stem at around 10 d, and fine roots began to appear at around 15 d. The root growth tended to be stable after one month, and the number of fibrous roots on the buds induced under the -N treatment was lower than those induced under the F treatment (Figures 4 and 5). Different concentrations of IBA and NAA had significant effects on the rooting of *A. mangium* and *A. auriculiformis*. For the rooting culture of buds induced under the -N treatment, the combined concentrations of 0.5 mg·L$^{-1}$ NAA and 1.0–2.0 mg·L$^{-1}$ IBA was more favorable for the rooting of *A. mangium* under nitrogen-deficient culture. Treatment 8 (2.0 mg·L$^{-1}$ IBA + 0.5 mg·L$^{-1}$ NAA) led to a significantly higher rooting rate than the other treatments, with the highest rooting rate reaching 77.8% (Table 2). When the concentration of NAA was 0.1 mg·L$^{-1}$, the rooting rate of *A. auriculiformis* gradually increased with the increase in the IBA concentration. Treatment 4 (2.0 mg·L$^{-1}$ IBA + 0.1 mg·L$^{-1}$ NAA) achieved the highest rooting rate, which was 88.9% (Table 2). For the rooting culture of buds induced under the F treatment, the rooting rates of the *A. mangium* buds under treatment 5 (0.5 mg·L$^{-1}$ IBA + 0.5 mg·L$^{-1}$ NAA) and treatment 12 (2.0 mg·L$^{-1}$ IBA + 1.0 mg·L$^{-1}$ NAA) were significantly higher than those of the other treatments. The rooting rate under treatment 12 was the highest, reaching 80.0% (Table 2). The rooting rate of the *A. auriculiformis* buds under treatment 3 (1.5 mg·L$^{-1}$ IBA + 0.1 mg·L$^{-1}$ NAA) was significantly higher than those of the other treatments, with the highest rooting rate reaching 90.0% (Table 2). The seedlings of *A. mangium* and *A. auriculiformis* rooted in the nitrogen-deficient media exhibited green leaves, robust growth, and no symptoms of nitrogen deficiency. Additionally, there was no significant difference compared to the seedlings rooted in the complete nutrient media (Figures 4 and 5).

**Table 2.** The rooting rates of induced *A. mangium* and *A. auriculiformis* buds.

| No. | Treatment | | *A. mangium* | | *A. auriculiformis* | |
|---|---|---|---|---|---|---|
| | Concentration of IBA (mg/L) | Concentration of NAA (mg/L) | Rooting Rate (-N) | Rooting Rate (F) | Rooting Rate (-N) | Rooting Rate (F) |
| 1 | 0.5 | 0.1 | 33.3 cd | 41.1 bcd | 66.7 bcd | 58.9 cd |
| 2 | 1.0 | | 24.4 cd | 34.4 cd | 76.7 ab | 75.6 abc |
| 3 | 1.5 | | 28.9 cd | 52.2 bc | 80.0 ab | 90.0 a |
| 4 | 2.0 | | 31.1 cd | 33.3 acd | 88.9 a | 64.5 bcd |
| 5 | 0.5 | 0.5 | 21.1 d | 75.6 a | 54.4 cd | 52.2 d |
| 6 | 1.0 | | 58.9 b | 31.1 d | 50.0 d | 66.7 bcd |
| 7 | 1.5 | | 66.7 ab | 47.8 bcd | 54.5 cd | 60.0 cd |
| 8 | 2.0 | | 77.8 a | 56.7 b | 64.4 bcd | 51.1 d |
| 9 | 0.5 | 1.0 | 37.8 c | 47.8 bcd | 57.8 cd | 54.4 d |
| 10 | 1.0 | | 34.4 cd | 57.8 b | 66.7 bcd | 61.1 cd |
| 11 | 1.5 | | 34.4 cd | 46.7 bcd | 82.2 a | 65.5 bcd |
| 12 | 2.0 | | 38.9 c | 80.0 a | 71.1 bc | 78.9 abc |

Different lower-case letters indicate significant differences at $p < 0.05$. Data are means ± standard errors.

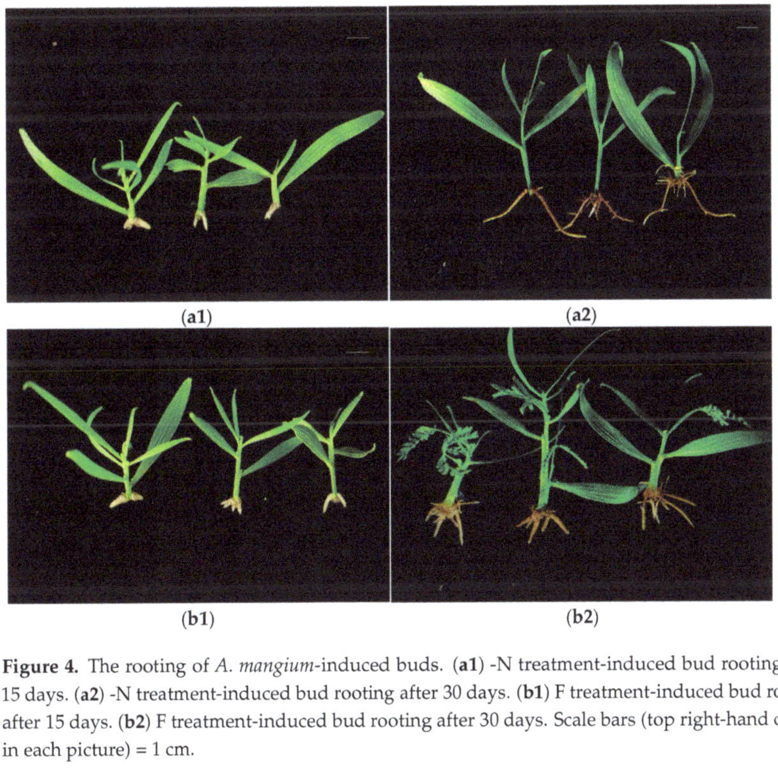

**Figure 4.** The rooting of *A. mangium*-induced buds. (**a1**) -N treatment-induced bud rooting after 15 days. (**a2**) -N treatment-induced bud rooting after 30 days. (**b1**) F treatment-induced bud rooting after 15 days. (**b2**) F treatment-induced bud rooting after 30 days. Scale bars (top right-hand corner in each picture) = 1 cm.

**Figure 5.** The rooting of *A. auriculiformis*-induced buds. (**c1**) -N treatment-induced bud rooting after 15 days. (**c2**) -N treatment-induced bud rooting after 30 days. (**d1**) F treatment-induced bud rooting after 15 days. (**d2**) F treatment-induced bud rooting after 30 days. Scale bars (top right-hand corner in each picture) = 1 cm.

## 3. Discussion

In this study, the effects of nitrogen deficiency on the bud induction of *A. mangium* and *A. auriculiformis* were studied. The research results show that nitrogen deficiency reduced the stem length and leaf number of the induced buds, but the induced buds grew robustly with dark green leaves and did not show obvious deficiency symptoms. *A. mangium* had the highest bud induction rates when treated with 1/4MS (-N) + 1.0 mg·$L^{-1}$ 6-BA and 1/8MS (F) + 0.5 mg·$L^{-1}$ 6-BA, reaching 72.6% and 91.9%, respectively, with no significant difference between the treatments with highest bud induction rates. *A. auriculiformis* had the highest bud induction rates when treated with 1/4MS (-N) + 1.0 mg·$L^{-1}$ 6-BA and 1/8MS (F) + 1.5 mg·$L^{-1}$ 6-BA, reaching 100.0% and 94.1%, respectively, with no significant difference between the treatments with highest bud induction rates. The highest rooting rates of the *A. mangium* buds induced under the -N and F treatments were 77.8% and 80.0%, respectively, with no significant difference between them. The highest rooting rates of the *A. auriculiformis* buds induced under the -N and F treatments were 88.9% and 90.0%, respectively, with no significant difference between them. Under the nitrogen-deficiency conditions, *A. mangium* and *A. auriculiformis* buds could be induced and grow normally, indicating that nitrogen is not essential during bud induction.

Different plants need different nutrition for growth and development during tissue culture. This study showed that *A. mangium* and *A. auriculiformis* buds could be induced in both nitrogen-deficient and complete nutrient media, indicating that nitrogen had little influence on bud induction. The degree of explant browning is directly related to the synthesis of phenolic substances [24]. An excessive concentration of inorganic salts in the medium may cause explants to produce a large amount of phenolic substances, with then undergo oxidation reactions, leading to browning [25,26]. This study showed that the inducted buds of *A. mangium* and *A. auriculiformis* were affected by the concentration of MS medium; the higher concentration ratio, the lower the bud induction rate and the higher the browning rate. This indicates that the bud induction of *Acacia* was very sensitive to high concentrations of the complete nutrient medium, which may have been due to the inhibition of germination caused by the penetration and toxicity of high concentrations of inorganic salts. In the nitrogen-deficient media, the browning rates of *A. mangium* and *A. auriculiformis* were lower than those in the complete nutrient media, and the bud induction rates in the MS (-N) and 1/2MS (-N) media were significantly higher than those in the corresponding complete nutrient media. This indicates that the lack of ammonium nitrate and potassium nitrate in the medium reduced the content of inorganic salts and inhibited the oxidation of phenolic substances in the explants. Therefore, the browning of the explants decreased, mortality was reduced, and the bud induction rates significantly increased.

The appropriate concentration of growth regulators can promote the induction and growth of buds. Qiu et al. found that adding appropriate plant hormones could effectively promote cell growth and alleviate browning [27]; Feng et al. found that the browning rate of explants increased with increases in the 6-BA concentration [28]. In this study, there were no significant differences in the browning rate, contamination rate, or bud induction rate of *A. mangium* and *A. auriculiformis* when the concentration of 6-BA was in the range of 0.5–1.5 mg·L$^{-1}$. This indicates that different plants have different tolerance degrees to the concentration of 6-BA, but 6-BA had no effect on the bud induction of *A. mangium* or *A. auriculiformis*. Growth regulators are crucial for the rooting of seedlings. Different concentrations of IBA and NAA have been used in the rooting of *Acacia* in vitro. We compared the buds induced in nitrogen-deficient and complete nutrient media by culturing them in rooting media with certain concentrations of IBA and NAA. The rooting rates ranged from 20% to 90%, depending on the combined concentrations of the growth media. *A. mangium* and *A. auriculiformis* had higher rooting rates in both the nitrogen-deficient and complete nutrient media with suitable concentrations of IBA and NAA, which further shows that nitrogen is not essential for *Acacia* tissue culture.

Under nitrogen stress, plants will show certain resistance. A study showed that in response to nitrogen-deficiency stress, *Hordeum vulgare* var. coeleste L. adapted by changing the activities of POD, CAT, and APX. These changes reduced the damage caused by nitrogen-deficiency stress, together playing a key role in enabling *Hordeum vulgare* var. coeleste L. to adapt to such stress [29]. Huo et al. found that when apple leaves and roots were under nitrogen-deficient conditions, the *MDATG10* gene was overexpressed, which in turn regulated the absorption and transformation of nitrogen, reduced the impact of nitrogen stress on chloroplasts, and enabled the plants to grow normally [30]. This study found that buds of *A. mangium* and *A. auriculiformis* could be effectively induced when a certain concentration of 6-BA was added to the media. Additionally, the bud induction of *A. auriculiformis* increased the demand for essential nutrients under the conditions of nitrogen deficiency. We speculate that under nitrogen-deficiency stress, explants of *A. auriculiformis* require more nutrients and stimulation, while *A. mangium* explants are more affected by 6-BA. *A. mangium* and *A. auriculiformis* adjust the absorption and transformation of nitrogen through their own physiological adaptation to nitrogen-deficient environments to meet their normal growth needs.

In summary, *A. mangium* and *A. auriculiformis* tissue culture seedlings were able to achieve normal growth and metabolism under nitrogen-deficient conditions. The buds induced under the nitrogen-deficiency treatment did not show obvious symptoms of nitrogen deficiency, and their growth status was not very different from those induced under the complete nutrient treatment. Additionally, the buds induced under nitrogen deficiency could root normally, indicating that nitrogen is not essential in the tissue culture of *A. mangium* and *A. auriculiformis*.

This study conducted a rooting experiment on buds induced in nitrogen-deficient medium, elucidating important condition requirements for growth and development in subsequent seedling refining and transplanting processes. It also successfully solved the challenge posed by the difficulty of purchasing potassium nitrate and ammonium nitrate for tissue culture production of *A. mangium* and *A. auriculiformis,* while providing a new idea for the sustainable development of tissue culture technology in the future.

## 4. Materials and Methods

### 4.1. Locations and Materials

The experimental materials were collected from the cutting orchard of *A. mangium* and *A. auriculiformis* at the *Acacia* Breeding Base in Shadui Town, Jiangmen City, Guangdong Province. The selected new shoots of the current year of *A. mangium* and *A. auriculiformis* exhibited good growth and were free from pests and diseases.

### 4.2. Methods

#### 4.2.1. Pretreatment of Explants

The selected new shoots of *A. mangium* and *A. auriculiformis* were soaked in washing powder dilution for 30 min and then rinsed. Then, a soft-bristled brush dipped in washing powder dilution was used to clean the surfaces of the branches. The stems with leaves were cut into 3–5 cm segments, retaining 1/3 of the original leaves and containing full axillary buds, to be used as explants. The pruned stems were rinsed with running water for more than 1 h, and then completely immersed in 0.8 g·L$^{-1}$ carbendazim solution for disinfection for 3 min [31,32]. Then, the explants were rinsed with tap water, and filter paper was used to dry the surface of the explants before inoculation.

#### 4.2.2. Screening of Nitrogen-Deficient Medium

The explants were inoculated on MS (-N), 1/2MS (-N), 1/4MS (-N), and 1/8MS (-N), where -N indicates that the culture medium lacked the main nitrogen sources ammonium nitrate and potassium nitrate. Then, different concentrations of 6-BA (0.5 mg·L$^{-1}$, 1.0 mg·L$^{-1}$, and 1.5 mg·L$^{-1}$) were added to the media, followed by 5 g·L$^{-1}$ AGAR (AisaBio, Gel strength > 1400 g/cm$^2$) and 0.2 g·L$^{-1}$ chlorothalonil. Each treatment was controlled with the same proportion of complete nutrient media, marked as F (the treatment media are shown in Table 3). Each treatment was inoculated with three explants per bottle, for a total of 15 bottles, and was repeated three times. The growth status of the explants was observed and recorded. After 40 days of culture, the following bud induction-related indicators were recorded: bud induction rate, contamination rate, browning rate, bud length, and leaf number.

**Table 3.** Treatment media used in this study.

| No. | Basic Medium | Concentration of 6-BA (mg/L) |
|---|---|---|
| 1 | MS(-N), MS(F) | 0.5 |
| 2 | MS(-N), MS(F) | 1.0 |
| 3 | MS(-N), MS(F) | 1.5 |
| 4 | 1/2MS(-N), 1/2MS(F) | 0.5 |
| 5 | 1/2MS(-N), 1/2MS(F) | 1.0 |
| 6 | 1/2MS(-N), 1/2MS(F) | 1.5 |
| 7 | 1/4MS(-N), 1/4MS(F) | 0.5 |
| 8 | 1/4MS(-N), 1/4MS(F) | 1.0 |
| 9 | 1/4MS(-N), 1/4MS(F) | 1.5 |
| 10 | 1/8MS(-N), 1/8MS(F) | 0.5 |
| 11 | 1/8MS(-N), 1/8MS(F) | 1.0 |
| 12 | 1/8MS(-N), 1/8MS(F) | 1.5 |

#### 4.2.3. Rooting Culture

The buds induced in the nitrogen-deficient medium and complete nutrient medium were inoculated into rooting medium for rooting culture. The rooting medium consisted of MS medium with different concentrations of IBA (0.5 mg·L$^{-1}$, 1.0 mg·L$^{-1}$, 1.5 mg·L$^{-1}$, and 2.0 mg·L$^{-1}$) and NAA (0.1 mg·L$^{-1}$, 0.5 mg·L$^{-1}$, and 1.0 mg·L$^{-1}$). A total of 12 treatments were set up. Each treatment was inoculated with three strains per bottle, for a total of 15 bottles, and repeated three times. After 30 days, rooting was observed, and the rate of rooting was calculated.

### 4.3. Culture Conditions

The media did not require high-temperature autoclave sterilization, and the operating environment was not strictly sterile. The culture conditions were as follows: pH of 5.8 ± 0.2, temperature of (25 ± 2) °C, 12 h of light per day, and light intensity of 2500 Lux.

### 4.4. Data Processing

Excel was used to organize the collected data, and the SPSS 26.0 statistical analysis software was used for analysis of variance ($p < 0.05$).

**Author Contributions:** Conceptualization, L.H.; methodology, Y.L.; software, L.S. and Y.L.; formal analysis, L.S. and Y.L.; investigation, Y.L.; resources, L.H.; data curation, L.S. and Y.L.; writing—original draft preparation, L.S.; writing—review and editing, L.S. and L.H.; project administration, L.H.; funding acquisition, L.H. All authors have read and agreed to the published version of the manuscript.

**Funding:** This research was funded by the Fundamental Research Funds of CAF, grant number No. CAFYBB2017 MB009.

**Data Availability Statement:** Data sharing is not applicable to this article.

**Conflicts of Interest:** The authors declare no conflicts of interest. The funders had no role in the design of the study; in the collection, analyses, or interpretation of data; in the writing of the manuscript; or in the decision to publish the results.

## References

1. Zhang, R.P.; Zhan, N.; Huang, L.J.; Zeng, B.S. Control Pollination of Acacia mangium and Acacia auriculiformis. *Mol. Plant Breed.* **2020**, *18*, 5877–5882.
2. Lu, D.D.; Wu, B.G.; Wang, X.Q.; Zheng, L.; Yonggang, J. A General Review on the Research Development of *Acacia* spp. *J. Fujian Coll. For.* **2004**, *24*, 92–96.
3. He, B.; Jia, L.M.; Jin, D.G.; Qin, W.M. Studies on Soil Fertility Change in *Acacia mangium* Plantation in Nanning, Guangxi. *Sci. Silvae Sin.* **2007**, *43*, 10–16.
4. Yi, M.; Huang, L.J.; Chen, Z.X.; Li, J. Cutting Propagation of *Acacia mangium*. *For. Res.* **2010**, *23*, 910–913.
5. Yi, M.; Huang, L.J.; Chen, Z.X.; Li, J. Techniques for Rooting Cuttings of *Acacia auriculiformis*. *Acta Agric. Univ. Jiangxiensis* **2011**, *33*, 84–89.
6. Zhan, N.; Huang, L.J. Conditions for in vitro Germination and Testing Method for Pollen Viability of *Acacia auriculiformis*. *Sci. Silvae Sin.* **2016**, *52*, 67–73.
7. Cai, L.; Wang, Y.H.; Wu, M.; Su, X. Study on Tissue Culture of Five *Acacia* Species. *Guangxi For. Sci.* **2003**, *1*, 24–26.
8. Hu, F.; Shi, Q.; Hang, L.J. Induction of Adventitious Roots During Tissue Culture of *Acacia mangium* and *A. auriculiformis* Elite Trees. *Nat. Sci. Ed.* **2015**, *39*, 57–62.
9. Tao, A.L.; Cao, D.J.; Hua, F.; Li, S.N.; Yang, M.L. Advances in Plant Tissue Culture Technology. *Nat. Sci. Ed.* **2018**, *15*, 31–35.
10. Zhan, N.; Huang, L.J. General Review on the Research Development of *Acacia* spp. in China. *Trop. For.* **2015**, *43*, 41–45.
11. Wang, J.X.; Wan, X.R.; Li, L.; Pan, R. Tissue Culture of *Acacia Confusa*. *Nat. Sci. Ed.* **2001**, *3*, 45–48+54.
12. Wang, H.; Huang, L.J.; Hu, F. An Efficient Tissue Culture Technique of *Acacia mangium* from 16-year old Plant. *Mol. Plant Breed.* **2016**, *14*, 986–996.
13. Wang, H.; Huang, L.J.; Hu, F. Efficient bud inducing and proliferation of 16-year-old *Acacia auriculiformis* plant. *J. South China Agric. Univ.* **2016**, *37*, 91–97.

14. Huang, L.J.; Wang, H.; Hu, F.; Shi, Q. Effects of Plant Growth Regulators on Axillary Bud Proliferation and Rooting of *Acacia melanoxylon* Elite Tree. *J. Trop. Subtrop. Bot.* **2015**, *23*, 669–674.
15. Shi, Q.; Hu, F.; Huang, L.J.; Zhan, N. An Efficient Axillary Bud Multiplication System of *Acacia mangium* × *A. auriculiformis*. *Sci. Silvae Sin.* **2014**, *50*, 55–60.
16. Wu, A.L.; He, B.; Wu, J.Y.; Zhou, X.J. Response Surface Methodology for Optimization of Adventitious Bud Induction and Proliferation Cultures of *A. podalyriifolia*. *Mol. Plant Breed.* **2020**, *18*, 7868–7878.
17. Coruzzi, G.; Bush, D.R. Nitrogen and Carbon Nutrient and Metabolite Signaling in Plants. *Plant Physiol.* **2001**, *125*, 61–64. [CrossRef]
18. Ma, J.H.; Qi, Y.Q.; Ren, Q.F.; Liu, F.; Ou, M.Z. Effects of Nitrogen on the Growth and Polysaccharides Accumulation of *Bletilla striata*. *Plant Physiol. J.* **2022**, *58*, 331–338.
19. Ye, Y.Q.; Luo, H.Y.; Li, M.; Liu, X.X.; Cao, G.Q.; Xu, S.S. Effects of Nitrogen Forms on Lateral Roots Development and Photosynthetic Characteristics in Leaves of *Cunninghamia lanceolata* Seedlings. *Acta Bot. Boreali-Occident. Sin.* **2018**, *38*, 2036–2044.
20. Qiu, Z.F.; Zeng, B.S.; Fan, C.J. Establishment of Regeneration System for Stem Segments of Superior Clone SR17 in *Acacia melanoxylon*. *Mol. Plant Breed.* **2018**, *16*, 5023–5028.
21. Wang, Y.; Yao, R.L. Effects of Different Nitrogen Forms and Ratios on Growth of Tissue Cultured Seedlings in *Pinus massoniana*. *J. Cent. South Univ. For. Technol.* **2021**, *41*, 18–24+71.
22. Hu, F.; Shi, Q.; Huang, L.J. Tissue Culture System of Axillary Proliferation of *Acacia crassicarpa*. *Bull. Bot. Res.* **2015**, *35*, 179–184.
23. *GB18218-2018;* Identification of Major Hazard Installations for Hazardous Chemicals. State Administration for Market Regulation of the People's Republic of China, Standardization Administration of the People's Republic of China: Beijing, China, 2018.
24. Feng, D.D.; Wang, J.; Chen, J.P. Research Progress of Browning in the Plant Tissue Culture. *Acta Agric. Zhejiangensis* **2015**, *27*, 1108–1116.
25. Mei, X.G.; Dong, Y.L.; Pan, X.W. Studies of Avoiding Browning in Tissue Subculture of *Taxus chinensis*. *Nat. Prod. Res. Dev.* **2001**, *4*, 8–11.
26. Li, D.J.; Zhang, J.X.; Wei, J.F.; Li, S.J. Relationship Between Culture Medium and Culture Condition and Avoiding Browning in Cell Culture of *Taxus*. *Plant Physiol. Commun.* **2005**, *1*, 95–98.
27. Qiu, X.F.; Zhu, D.; Zhang, Z.B.; Tu, Y.S.; You, H. Study on Callus Subculture and Darkening Inhibition of *Eucommia ulmoides*. *Foodence* **2007**, *8*, 307–309.
28. Feng, Y.; Qian, L.W.; Lin, Q.L. The Effect of Different Hormones on Explant Browning and Callus Browning in *Cyclocarya paliurus*. *Chin. Bull. Bot.* **2019**, *54*, 634–641.
29. Wei, Z.X.; Xu, Q.J.; Za, S.; Wang, Y.L.; Yuan, H.J. The Effects of Nitrogen Deficiency on the Growth and Physiological Characteristics of *Hulless Barley* Seedlings. *Barley Cereal Sci.* **2017**, *34*, 1–9.
30. Huo, L.; Guo, Z.; Wang, Q.; Cheng, L.; Jia, X.; Wang, P.; Gong, X.; Li, C.; Ma, F. Enhanced Autophagic Activity Improved the Root Growth and Nitrogen Utilization Ability of *Apple* Plants under Nitrogen Starvation. *Int. J. Mol. Sci.* **2021**, *22*, 8085. [CrossRef]
31. Huang, L.J.; Wang, H.; Shahid, M.Q.; Zeng, B. Chlorothalonil: An Effective Bacteriostatic Agent for bud induction of *Acacia auriculiformis* under open condition (non-axenic). *Plant Methods* **2019**, *15*, 5. [CrossRef]
32. Lu, Y.P.; Huang, L.J.; Wang, H. *Acacia mangium* × *A. Auriculiformis* micropropagation in a non-sterile environment. *Aust. For.* **2024**, *87*, 17–25.

plants MDPI

*Article*

# Changes in the Quality of *Idesia polycarpa* Maxim Fruits from Different Ecotypes During the Growth Process

Yi Yang [1,2,3,4,†], Chao Miao [1,2,3,†], Qiupeng Yuan [1,2,3], Wenwen Zhong [1,2,3], Zuwei Hu [1,2,3], Chen Chen [1,2,3], Zhen Liu [1,2,3], Yanmei Wang [1,2,3], Xiaodong Geng [1,2,3], Qifei Cai [1,2,3], Li Dai [1,2,3], Juan Wang [1,2,3], Yongyu Ren [1,2,3], Fangming Liu [1,2,3], Haifei Lu [4], Tailin Zhong [4,*] and Zhi Li [1,2,3,*]

1 College of Forestry, Henan Agricultural University, Zhengzhou 450046, China
2 National Forestry and Grassland Administration Key Laboratory for Central Plains Forest Resources Cultivation, Zhengzhou 450046, China
3 Henan Province Engineering Technology Research Center for *Idesia*, Zhengzhou 450046, China
4 College of Urban Construction, Zhejiang Shuren University, Shaoxing 312028, China
* Correspondence: garden168zhong@163.com (T.Z.); lizhi@henau.edu.cn (Z.L.)
† These authors contributed equally to this work.

**Abstract**

The goal of this study was to build an understanding of the quality of *Idesia polycarpa* fruit Maxim from different ecotypes and to identify the best cultivars, with a view to providing a reference and theoretical basis for the selection and cultivation of *I. polycarpa*. In this study, we systematically evaluated the fruit quality characteristics of five seed sources, namely SH, SG1, GG, HX, and SG2, at four developmental stages, M1-M4, through a principal component analysis, a correlation analysis, and a significance test. Comparisons between the ecotype yielded that GG was significantly better than the other ecotype in oil content (28.7%) and fresh weight per cluster (155.56 g), while HX exhibited higher SOD content (278.18 U/g) and soluble protein content (27.50 $mg \cdot g^{-1}$), suggesting a higher level of stress tolerance. The results of the correlation analysis showed that POD was significantly negatively correlated with oil content ($r = -0.633$) and SOD ($r = -0.617$) activities, indicating that the antioxidant enzyme system may affect oil accumulation. The results of the principal component analysis showed that the cumulative contribution of the first four principal components reached 89.72%, of which principal component 1 mainly reflected yield-related traits, and principal component 2 was significantly correlated with oil content and soluble protein. Through the evaluation and screening of the five ecotypes, we determined that GG can be utilized as a good single plant in the selection and improvement of new cultivars; our findings can provide theoretical support for the selection of good cultivars of *I. polycarpa* seed in the central region of Henan.

**Keywords:** *Idesia polycarpa*; fruit; quality; yield; seed source variation

## 1. Introduction

*Idesia polycarpa* Maxim belongs to Salicaceae; the plant is also known as oil grapes, water winter melon, mountain sycamore, etc., and is referred to as the "tree on the oil bank" [1]. China is a center for *I. polycarpa* distribution. Most provinces in China have a distribution of wild *I. polycarpa*; its wide distribution and adaptability mean that it is an excellent tree species for use in the afforestation and greening of deserted mountains [2]. The tree crown of *I. polycarpa* is of a layered type; the fruit is bright red in color after maturity, with a string-hanging sequence; the wood hardness is moderate; this species is adaptable, and can be used as an ornamental, ecological wood species in gardening and greening applications [3]. The fruit yield of *I. polycarpa* is very high, measuring more than 1.5 kg/m$^2$ [4]. The fruit of *I. polycarpa* can be used as a raw material in various applications: edible oil extraction [5]; linoleic acid extraction [6,7]; bio-lubricant production [8,9]; other aspects of preparation technologies. Developments in the *I. polycarpa* industry have laid a solid national foundation. The pulp oil content from the plant can be as high as 44.08% [10], with a seed oil content of 26.54%; accordingly, *I. polycarpa* is an important woody oilseed species in China [11].

Fruit morphologies and physiological and biochemical indices are fundamental parameters that must be determined in elucidating the mechanisms underlying fruit growth and development. In recent years, several studies have explored various aspects of *I. polycarpa*. For example, the oil content and fatty acid composition of *I. polycarpa* from five regions were analyzed by Wang et al. [12]. Their findings indicated that fruits from Henan Province exhibited the highest oil content, while oil derived from *I. polycarpa* had the highest linoleic acid content. Seedling growth indices of *I. polycarpa* of 12 provenances were measured by Jiang et al. [13], who determined that seedlings of the Fanyi provenance showed the fastest growth rate. According to Gong et al. [14], who investigated the fruit fatty acid profiles of *I. polycarpa* samples from 11 provenances, the fruits are rich in unsaturated fatty acids, with linoleic acid being the predominant component. Hernández et al. [15] showed that the fatty acid profile (especially oleic/linoleic acid ratio) and its genomic expression have a significant effect on oil stability. The chemical composition and antioxidant activity of the pulp oil from five places in China were analyzed by Zhang et al. [16]; meanwhile, Ye et al. [17] optimized the extraction process of polyphenols from the defatted fruit residue of S. sanguinis and their in vitro/in vivo antioxidant and whitening activities. The photosynthetic characteristics of *I. polycarpa* samples from four provenances were investigated by Xu et al. [18]; their results showed that the sample of Zhangjiajie provenance had the highest light adaptability, while that of Tokyo provenance had the widest range of net photosynthetic rate (Pn) and the highest Pn value. A quality evaluation of seven provenances of *I. polycarpa* fruits and their oils was carried out by Zou et al. [19], who found provenance to be a significant factor influencing quality. Experiments on *I. polycarpa* under different environmental conditions were performed by Xu et al. [20], with the results demonstrating strong resistance characteristics. The response mechanism of *I. polycarpa* under drought stress [21] and flood stress [22] was explored by Geng et al., revealing that moderate drought could promote seedling growth while showing some tolerance to flood stress. Research conducted by Tian et al. [23] on *I. polycarpa* under extreme high temperatures and high temperature–drought stress showed that high temperature–drought stress caused greater damage than high-temperature stress alone. In addition, significant progress has been made in research on the physiological mechanisms and cultivation management of *I. polycarpa*, an ecologically and economically valuable tree species. Feng et al. addressed disease issues in practical production by systematically identifying the pathogen responsible for stem canker in *I. polycarpa* for the first time and developing effective chemical control strategies, providing a scientific basis for integrated disease management [24]. In the field of root ecology, Li et al. evaluated the fine root morphology and rhizosphere environmental characteristics of dioecious *I. polycarpa*, revealing the potential influence of gender differentiation on root architecture and soil microenvironments, thereby laying the foundation for sex-specific cultivation practices [25]. Zhang et al. focused on the regulatory mechanisms of seed germination. Through transcriptome analysis, they elucidated the molecular network by which temperature drives seed germination through the coordinated regulation of plant hormone signaling and sugar metabolism pathways [26]. Regarding nutrient management, Wang et al. investigated the effects of varied phosphorus fertilizer ratios on the rhizosphere microbial community structure in seedlings. Their study confirmed that phosphorus levels significantly alter bacterial and fungal diversity, consequently impacting the rhizosphere microecological balance [27]. In research on sex determination mechanisms, Wang et al. integrated proteome and transcriptome data. By comparing mutant bisexual flowers with wild-type male flowers, they identified key regulatory genes and metabolic pathways involved in the sex conversion of *I. polycarpa*, offering molecular targets for the artificial regulation of sexual expression [28]. However, despite these valuable contributions, our understanding of fruit development characteristics across *I. polycarpa* cultivars remains incomplete. Further in-depth research is needed if we are to build a more comprehensive foundation of understanding for cultivar selection.

In light of the above-described context, in this study, we selected five provenances of *I. polycarpa* as experimental materials: SH, SG1, GG, HX, and SG2. By systematically comparing the dynamic changes in fruit growth and development among these provenances, a comprehensive analysis was conducted on the fruits' phenotypic traits and physiological–biochemical variations. The objectives were to scientifically evaluate the quality characteristics from *I. polycarpa* from different provenances, screen superior cultivars with excellent overall performances, and finally provide substantive references and theoretical underpinnings for the genetic improvement, breeding, and cultivation of *I. polycarpa*.

## 2. Results

### 2.1. Changes in Phenotypic Traits During Fruit Growth of *I. polycarpa* from Different Ecotypes

From June to September, fruit phenotypic changes were observed in the different *I. polycarpa* ecotypes used in this study (Tables 1 and 2). In terms of cluster length, the GG seed source had the longest cluster length in the M1-M4 development stages, which was about 105% higher than that of SG1, which had the shortest cluster length; SH and SG2 ecotype had middling lengths, but SG2 had longer lengths in the M1 stage, which was probably related to the rapid growth that took place in the early stage. In terms of cluster width, the differences among the four stages were small, mostly between 8 and 11 cm, but GG was significantly wider than the other ecotype in the M2-M4 stages, and the width of HX was the most stable. In terms of the cluster aspect ratio, the ratios of GG and SG2 were the highest, with the clusters being slender and long, while the ratio of SG1 was the lowest, with the clusters being thicker and shorter. In terms of the fresh weight of single clusters, the weight of GG was significantly higher than those of the other ecotypes, and the fresh weight of a single cluster of GG was higher than those of the other ecotypes. In terms of fresh weight per cluster, the weight of GG was significantly higher than those of the other ecotypes: it was 3.1 times higher than that of SH; HX had a peak at the M3 stage, but it dropped sharply by 46% at the M4 stage. In terms of the percentage of fresh weight per cluster, there were small differences between the five ecotypes in terms of the proportion of fresh weight of the fruit to the fresh weight of a single cluster; however, GG had the highest proportion at the M4 stage, indicating that it has higher economic value. In terms of the number of fruits in a single cluster, GG had the highest ratio; lower ratios were found to be associated with smaller clusters. The number of fruits for GG was much higher than those for the other ecotypes: it was 4.7 times higher than that of SG1, and the numbers of fruits in SH and SG2 were about the same in the M3 stage; however, the number of fruits in SG2 declined by 36% in the M4 stage, and the number of fruits in SH remained stable. The 1000-seed weight for SG1 was the highest among the five ecotypes, indicating good growth. The fruit shape index and the weight of single fruit showed small differences. For the fruit shape index and single-fruit weight, most fruit shape indexes ranged from 0.90 to 1.11, and fruit shapes were close to round or slightly longer; HX showed the best shape among all the ecotypes.

**Table 1.** Phenotypic traits during fruit growth of *I. polycarpa* from different ecotypes (first part).

| Seed Source | Developmental Stage | Cluster Length (cm) | Cluster Width (%) | Width/Length Ratio (%) | Fresh Weight per Cluster (g) | Fresh Weight of Fruit Stalks per Cluster (g) | Fruit Weight per Cluster (g) |
|---|---|---|---|---|---|---|---|
| SH | M1 | 21.73 ± 1.03Db | 8.10 ± 0.40Ba | 2.69 ± 0.11Ba | 22.46 ± 6.22Ca | 2.49 ± 0.71Cb | 19.97 ± 5.53Ca |
| | M2 | 27.30 ± 1.85Ba | 9.87 ± 1.40Aa | 2.83 ± 0.24ABa | 36.47 ± 7.46Ba | 4.36 ± 1.03Bab | 32.11 ± 6.47Ba |
| | M3 | 26.83 ± 0.95BCa | 9.20 ± 0.38Ba | 2.93 ± 0.22ABa | 49.35 ± 11.06Ca | 5.35 ± 0.65BCab | 44.01 ± 10.41Ca |
| | M4 | 29.07 ± 1.65Ba | 10.20 ± 0.29ABa | 2.85 ± 0.08Ba | 49.82 ± 8.31Ba | 5.93 ± 0.94Ba | 43.89 ± 7.56Ba |
| SG1 | M1 | 23.56 ± 1.20CDab | 8.37 ± 0.15Ba | 2.82 ± 0.17Ba | 30.50 ± 0.87BCa | 3.47 ± 0.20Ca | 27.03 ± 0.89BCa |
| | M2 | 21.33 ± 2.14Cb | 9.03 ± 0.78Aa | 2.36 ± 0.09Bb | 32.19 ± 5.52Ba | 2.96 ± 0.64Ba | 29.23 ± 4.89Ba |
| | M3 | 21.77 ± 1.91Cab | 8.47 ± 0.52Ba | 2.57 ± 0.13Bab | 42.85 ± 15.31Ca | 4.31 ± 1.70Ca | 38.54 ± 13.61Ca |
| | M4 | 27.20 ± 1.21Ba | 10.00 ± 0.23ABa | 2.72 ± 0.06Bab | 55.82 ± 3.85Ba | 6.01 ± 0.57Ba | 49.81 ± 3.29Ba |
| GG | M1 | 34.93 ± 0.49Ab | 9.70 ± 0.10Ab | 3.60 ± 0.06Aab | 63.33 ± 10.08Ac | 8.00 ± 1.47Ab | 55.33 ± 8.63Ac |
| | M2 | 37.17 ± 1.14Ab | 11.43 ± 0.35Aa | 3.25 ± 0.13Ab | 76.73 ± 2.45Abc | 7.12 ± 1.55ABb | 69.61 ± 2.02Abc |
| | M3 | 39.20 ± 2.69Aab | 11.40 ± 1.04Aa | 3.44 ± 0.10Ab | 100.05 ± 13.70ABb | 9.78 ± 1.75Ab | 90.26 ± 11.96ABb |
| | M4 | 43.87 ± 0.88Aa | 11.27 ± 0.92Aa | 3.90 ± 0.11Aa | 155.56 ± 7.70Aa | 15.97 ± 0.66Aa | 139.60 ± 7.23Aa |
| HX | M1 | 28.13 ± 2.28BCa | 8.17 ± 0.50Ba | 3.44 ± 0.09Aa | 53.43 ± 12.48ABb | 4.81 ± 1.05BCa | 48.62 ± 11.45ABb |
| | M2 | 27.43 ± 2.45Ba | 10.00 ± 0.95Aa | 2.75 ± 0.04ABa | 85.58 ± 21.92Aab | 6.80 ± 1.86ABa | 78.79 ± 20.09Aab |
| | M3 | 32.10 ± 1.31Ba | 10.33 ± 1.01ABa | 3.18 ± 0.39ABa | 110.01 ± 14.23Aa | 7.55 ± 0.57ABCa | 102.46 ± 13.75Aa |
| | M4 | 29.17 ± 2.06Ba | 9.07 ± 0.43Ba | 3.23 ± 0.28ABa | 56.77 ± 7.10Bb | 4.84 ± 0.31Ba | 51.94 ± 6.84Bb |
| SG2 | M1 | 31.07 ± 1.76ABa | 8.33 ± 0.43Ba | 3.73 ± 0.03Aa | 56.56 ± 4.52ABa | 7.82 ± 0.88ABa | 48.74 ± 3.65ABa |
| | M2 | 30.47 ± 0.23Ba | 10.13 ± 0.78Aa | 3.05 ± 0.25Aab | 63.91 ± 7.78ABa | 8.81 ± 0.95Aa | 55.10 ± 6.94ABa |
| | M3 | 26.10 ± 1.81BCa | 9.90 ± 0.15ABa | 2.64 ± 0.22Bb | 65.36 ± 4.13BCa | 8.58 ± 0.75ABa | 56.78 ± 3.66BCa |
| | M4 | 30.00 ± 2.67Ba | 9.70 ± 0.72ABa | 3.17 ± 0.51ABab | 48.46 ± 7.15Ba | 7.84 ± 1.68Ba | 40.62 ± 5.53Ba |

Note: Uppercase letters after data in the same column indicate significant differences ($\alpha = 0.05$) across ecotypes; lowercase letters indicate significant differences ($\alpha = 0.05$) across time. Data are presented as mean ± standard error, with error values rounded to two significant digits.

**Table 2.** Phenotypic traits during fruit growth of *I. polycarpa* from different ecotypes (part two).

| Seed Source | Developmental Stage | Number of Fruits per Cluster | Seed Weight per Thousand Grains (g) | Fruit Width (mm) | Fruit Length (mm) | Fruit Shape Index | Fruit Mass (g) |
|---|---|---|---|---|---|---|---|
| SH | M1 | 88 ± 23Ca | 0.30 ± 0.03Bab | 7.28 ± 0.14Ca | 7.95 ± 0.03Ba | 1.09 ± 0.02Aa | 0.25 ± 0.03Ba |
| | M2 | 141 ± 22BCa | 0.29 ± 0.00Bb | 7.23 ± 0.28Ba | 8.01 ± 0.15Ba | 1.11 ± 0.03Aa | 0.23 ± 0.02Ba |
| | M3 | 149 ± 29BCa | 0.34 ± 0.02Aab | 7.93 ± 0.42Ca | 8.51 ± 0.21Ba | 1.08 ± 0.04Aa | 0.30 ± 0.03Ba |
| | M4 | 156 ± 21Ba | 0.36 ± 0.00Ba | 7.89 ± 0.39Ca | 8.38 ± 0.33Ba | 1.06 ± 0.01Aa | 0.30 ± 0.04Ba |
| SG1 | M1 | 75 ± 3Ca | 0.40 ± 0.02Ab | 8.64 ± 0.18Ab | 8.67 ± 0.06Ab | 1.00 ± 0.01Ba | 0.38 ± 0.01Ab |
| | M2 | 71 ± 10Ca | 0.43 ± 0.02Aab | 9.18 ± 0.37Aab | 9.17 ± 0.30Aab | 1.00 ± 0.01Ba | 0.44 ± 0.05Aab |
| | M3 | 82 ± 25Ca | 0.37 ± 0.02Ab | 9.35 ± 0.34ABab | 9.08 ± 0.27ABab | 0.97 ± 0.01Ba | 0.45 ± 0.04Aab |
| | M4 | 96 ± 5.5Ba | 0.48 ± 0.00Aa | 9.82 ± 0.03ABa | 9.54 ± 0.00Aa | 0.97 ± 0.00Ba | 0.51 ± 0.00Aa |
| GG | M1 | 228 ± 33Ab | 0.36 ± 0.00ABb | 7.84 ± 0.19Bb | 7.25 ± 0.11Cc | 0.93 ± 0.02CDa | 0.24 ± 0.02Bb |
| | M2 | 263 ± 15Ab | 0.40 ± 0.00Aa | 7.93 ± 0.17Bb | 7.34 ± 0.11Cc | 0.93 ± 0.01Ca | 0.26 ± 0.02Bb |
| | M3 | 319 ± 41Ab | 0.37 ± 0.01Ab | 8.14 ± 0.08Cb | 7.71 ± 0.05Cb | 0.95 ± 0.00Ba | 0.29 ± 0.01Bb |
| | M4 | 449 ± 27Aa | 0.38 ± 0.01Bb | 8.64 ± 0.08Ca | 8.17 ± 0.12Ba | 0.94 ± 0.00Ca | 0.35 ± 0.02Ba |
| HX | M1 | 139 ± 34BCa | 0.38 ± 0.02Aa | 9.11 ± 0.23Aa | 8.76 ± 0.14Ab | 0.96 ± 0.01BCab | 0.38 ± 0.03Ab |
| | M2 | 199 ± 48ABa | 0.39 ± 0.01Aa | 9.04 ± 0.20Aa | 8.64 ± 0.12Ab | 0.96 ± 0.01BCab | 0.39 ± 0.02Ab |
| | M3 | 205 ± 33Ba | 0.39 ± 0.02Aa | 9.97 ± 0.11Aa | 9.38 ± 0.12Aab | 0.94 ± 0.00Bb | 0.51 ± 0.02Aab |
| | M4 | 115 ± 26Ba | 0.37 ± 0.04Ba | 9.98 ± 0.52Aa | 9.68 ± 0.45Aa | 0.97 ± 0.01Ba | 0.53 ± 0.07Aa |
| SG2 | M1 | 208 ± 18ABa | 0.32 ± 0.01Ba | 7.89 ± 0.06Bb | 7.18 ± 0.11Cb | 0.91 ± 0.01Da | 0.24 ± 0.01Bb |
| | M2 | 198 ± 27ABa | 0.33 ± 0.02Ba | 8.03 ± 0.14Bab | 7.41 ± 0.07Cab | 0.92 ± 0.01Ca | 0.27 ± 0.01Bb |
| | M3 | 185 ± 24Ba | 0.24 ± 0.03Bb | 8.50 ± 0.29BCab | 7.69 ± 0.27Cab | 0.90 ± 0.01Ba | 0.31 ± 0.04Bab |
| | M4 | 118 ± 9Bb | 0.34 ± 0.02Ba | 8.83 ± 0.35BCa | 7.99 ± 0.29Ba | 0.90 ± 0.00Da | 0.38 ± 0.03Ba |

Note: Uppercase letters after data in the same column indicate significant differences ($\alpha = 0.05$) across ecotypes; lowercase letters indicate significant differences $\alpha$ = 0.05) across time. Data are presented as mean ± standard error, with the error value for the number of fruits per cluster rounded to the nearest whole number, and the error values for the remaining indicators retained to two significant figures.

### 2.2. Changes in Oil and Water Content of Fruits from Different Ecotypes of *I. polycarpa*

Changes in oil and water content of *I. polycarpa* fruits from different ecotypes from June to September (Table 3). The water content of *I. polycarpa* fruits of different ecotypes decreased progressively with the growth stage. For example, the water content of the SG1 seed source decreased from 69% in stage M1 to 59% in stage M4, which was a significant decrease of 15%. In the same developmental stage, there were significant differences between ecotypes, and the water content of seeds from SG1 was significantly higher than those from SH and SG2 at the M1 stage, indicating that its early water metabolism was stronger. Oil content showed an increasing trend with the development stages, e.g., the oil content of seeds from GG increased from 15% at the M1 stage to 29% at the M4 stage, reflecting a significant increase of 93.3%. The differences between the ecotypes were particularly evident at the later stages. The oil content rate of seeds from GG was significantly higher than those from SG1 and HX at the M4 stage, indicating that the GG seed source was most capable of oil accumulation, followed by SH and SG2.

### 2.3. Changes of Antioxidant Enzymes in Fruit *I. polycarpa* from Different Ecotypes

From June to September, changes were observed in the antioxidant enzymes in *I. polycarpa* fruits from different ecotypes (Table 3). POD activity was higher in the early stages of development and decreased significantly in the later stages, showing an inverted V-shaped development trend. For example, POD activity in HX, SG1, GG, SG2, and SH decreased significantly from the M1 stage to the M4 stage, with significant decreases of 96.7%, 94.7%, 93.6%, 90.8%, and 86.7%. The POD activity of SH was 392.95 U/g at the M3 stage, which was significantly higher than that of the other ecotype. SOD activity showed an opposite trend, with SH SOD activity increasing from 238.50 U/g at the M1 stage to 337.37 U/g at the M4 stage; this was significantly higher than that of the other ecotype, with a significant increase of 41.45%. In the comparisons between the ecotypes, the SOD activity of the GG seed source was 175.13 U/g at the M2 stage, which was significantly lower than that of the other ecotype.

### 2.4. Changes in Fruit Content of *I. polycarpa* from Different Ecotypes

From June to September, changes were observed in the fruit content of *I. polycarpa* from different ecotypes (Table 3). Soluble sugars showed a significant decrease from the M3 stage to the M4 stage. HX showed a significant decrease from 5.86 $mg \cdot g^{-1}$ at the M1 stage to 2.40 $mg \cdot g^{-1}$, with a decrease of 58.99%, which was the sharpest decline observed among the ecotypes. This was followed by a significant decrease from 7.53 $mg \cdot g^{-1}$ to 3.54 $mg \cdot g^{-1}$ in the M1 stage in GG, with a decrease of 52.97%. SG1 and SG2 showed significant changes throughout the four stages. Soluble protein, on the other hand, showed an increasing trend, with the highest accumulation of 118.69% in the SH seed source, followed by accumulations of 90.63% and 71.95% in SG2 and GG, respectively. The M3 and M4 stages of the SG2 seed source were 25.79 $mg \cdot g^{-1}$ and 31.88 $mg \cdot g^{-1}$, respectively, which were significantly higher than those of the other ecotypes.

**Table 3.** Indexes of oil content, water content, inclusions, and antioxidant enzymes during the growth of fruits from different ecotypes of *I. polycarpa*.

| Seed Source | Developmental Stage | Water Content (%) | Oil Content (%) | POD (U/g) | SOD (U/g) | Soluble Protein ($mg \cdot g^{-1}$) | Soluble Protein ($mg \cdot g^{-1}$) |
|---|---|---|---|---|---|---|---|
| SH | M1 | 62.7 ± 3.5Aab | 9.1 ± 0.1Bc | 1605.81 ± 337.60ABa | 238.50 ± 28.00ABb | 5.48 ± 0.13Ba | 12.16 ± 1.71Ad |
| | M2 | 63.3 ± 1.2Aa | 19.4 ± 1.1BCb | 1684.40 ± 82.83Aa | 276.88 ± 5.65Ab | 5.78 ± 0.14Ba | 15.86 ± 0.433Ac |
| | M3 | 61.3 ± 0.3Aab | 24.7 ± 1.1Bab | 392.95 ± 42.56Ab | 291.99 ± 13.25Aab | 6.67 ± 0.70Aa | 21.61 ± 0.64Bb |
| | M4 | 56.7 ± 0.7Bb | 28.0 ± 3.7Ba | 214.23 ± 5.70Ab | 337.37 ± 6.50Aa | 2.72 ± 0.13BCb | 26.60 ± 0.04BCa |
| SG1 | M1 | 68.6 ± 0.1Aa | 9.3 ± 0.5Bc | 2267.21 ± 69.84Aa | 224.44 ± 4.64Bb | 7.57 ± 0.37Ad | 16.57 ± 2.50Ab |
| | M2 | 62.2 ± 1.0ABbc | 16.0 ± 0.8Cb | 1300.60 ± 273.35Ab | 245.31 ± 15.90Bb | 6.47 ± 0.10Bc | 15.97 ± 1.11Ab |
| | M3 | 62.8 ± 1.5Ab | 19.3 ± 0.8Cab | 290.46 ± 14.30Ac | 253.29 ± 20.16Aab | 7.02 ± 0.17Ab | 20.33 ± 0.25BCab |
| | M4 | 59.2 ± 0.8Ac | 19.8 ± 1.8Ca | 119.49 ± 11.11Bc | 291.12 ± 6.42BCa | 3.84 ± 0.17Aa | 21.49 ± 0.56Da |
| GG | M1 | 66.2 ± 1.6Aa | 15.5 ± 0.7Ad | 1981.07 ± 47.30AaB | 223.42 ± 5.05Bc | 7.53 ± 0.18Aa | 14.49 ± 0.69Ac |
| | M2 | 63.2 ± 0.7Aab | 19.6 ± 0.7ABcC | 1646.45 ± 613.70Aa | 175.13 ± 8.24Dd | 6.97 ± 0.41Ba | 17.23 ± 0.96Abc |
| | M3 | 61.6 ± 0.2Ab | 21.7 ± 0.4Cb | 325.64 ± 95.32Ab | 289.93 ± 6.86Ab | 7.09 ± 0.23AAa | 18.13 ± 1.27Cb |
| | M4 | 58.3 ± 0.8ABc | 28.7 ± 0.6Ba | 125.93 ± 4.28Bb | 315.53 ± 5.96ABa | 3.54 ± 0.02Ab | 24.91 ± 0.68Ca |
| HX | M1 | 66.2 ± 0.4Aa | 15.6 ± 0.1Ac | 1820.46 ± 391.1ABa | 279.68 ± 1.17Aa | 5.86 ± 0.80Ba | 17.13 ± 1.03Ad |
| | M2 | 63.3 ± 0.3Ab | 23.9 ± 1.9Aab | 1591.15 ± 256.69Aa | 222.48 ± 7.97BCb | 6.07 ± 0.54Ba | 11.22 ± 0.06Bc |
| | M3 | 61.7 ± 0.2Ac | 25.6 ± 0.9Ba | 276.27 ± 7.02Ab | 287.87 ± 12.11Aa | 6.28 ± 0.29Aa | 19.79 ± 0.33BCb |
| | M4 | 59.3 ± 0.7Ad | 20.5 ± 0.9Cb | 59.26 ± 8.55Cb | 278.18 ± 7.80Ca | 2.40 ± 0.22Cb | 27.50 ± 0.09Ba |
| SG2 | M1 | 63.5 ± 0.6Aa | 15.1 ± 0.7Ac | 1342.72 ± 177.45Ba | 242.55 ± 6.32ABab | 5.93 ± 0.10Bc | 16.72 ± 0.28Ac |
| | M2 | 60.6 ± 1.0Bb | 20.6 ± 1.1ABb | 1433.07 ± 333.15Aa | 211.56 ± 4.60Cb | 9.34 ± 0.65Aa | 18.51 ± 1.21Ac |
| | M3 | 61.2 ± 0.6Aab | 24.0 ± 1.3Bab | 278.85 ± 16.65Ab | 276.53 ± 8.07Aa | 7.40 ± 0.18Ab | 25.79 ± 1.15Ab |
| | M4 | 58.7 ± 0.9ABb | 27.3 ± 1.3Ba | 123.13 ± 33.82Bb | 280.24 ± 19.40Ca | 2.96 ± 0.21Bd | 31.88 ± 0.82Aa |

Note: Uppercase letters after data in the same column indicate significant differences ($\alpha = 0.05$) across ecotypes; lowercase letters indicate significant differences ($\alpha = 0.05$) across time. The data are expressed as mean ± standard error, with the error values for water content and oil content retained in one significant digit; the error values for the remaining indicators are retained to two significant digits.

## 2.5. Correlation Analysis of Fruit Quality Indexes of *I. polycarpa*

Correlations among 19 traits of *I. polycarpa* fruits were investigated using Pearson correlation analysis (Figure 1). From the results, significant positive and negative correlations can be observed among several traits. Oil content and water content (r = −0.630) showed a highly significant negative correlations, and it is possible that oil accumulation came at the expense of water consumption. In addition, oil content was significantly and positively correlated with the traits of cluster width (r = 0.481), number of grains (r = 0.401), and fresh fruit weight per cluster (r = 0.522); moreover, it is possible that fruits with high oil content generally tended to have wider clusters, more grains, and higher fresh weight per cluster. Cluster length was found to be significantly and positively correlated with cluster width (r = 0.634); both were highly and significantly positively correlated with grain number (r = 0.848, 0.678) and the fresh weight of a single stalk (r = 0.802, 0.618), suggesting that increased cluster morphology may promote seediness and biomass. However, fruit shape index was significantly and positively correlated with cluster length (r = 0.738) and significantly and negatively correlated with fruit transverse and longitudinal diameter (r = −0.404, 0.413). The thousand-grain weight of the seeds was significantly and positively correlated with the fruits' transverse diameters and the single-fruit weight (r = 0.490, 0.512). POD activity was significantly positively correlated with water content (r = 0.654), and significantly negatively correlated with the oil content and the fresh weight of single fruit cluster (r = −0.633, −0.315); these results suggest that POD activity may increase in environments with higher water content, and its activity may be inhibited in environments with higher oil content. SOD activity showed a highly significant negative correlation with POD (r = −0.617), which could be antagonistic between POD and SOD. Soluble proteins were significantly and positively correlated with SOD (r = 0.515), suggesting a synergistic effect between the antioxidant system and proteins. Soluble sugars were significantly and positively correlated with POD activity (r = 0.490), but showed a highly significant negative correlation with SOD and soluble proteins (r = −0.563, −0.521), suggesting that sugar accumulation may be dependent on the antioxidant part of the system. Meanwhile, soluble sugar showed a significant negative correlation with fruit longitudinal diameter and single-fruit weight (r = −0.359, −0.317), suggesting that higher sugar may inhibit fruits' longitudinal development.

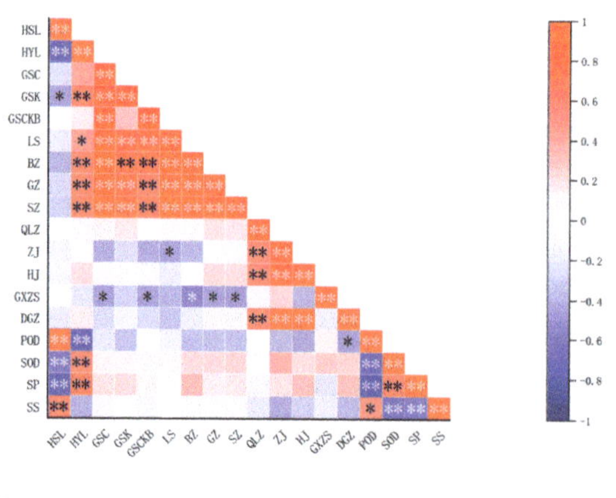


**Figure 1.** Correlation analysis of fruit quality indexes of different ecotypes *I. polycarpa*. Water content is denoted as HSL, oil content is denoted as HYL, cluster length is denoted as GSC, cluster width is

denoted as GSK, cluster aspect ratio is denoted as GSCKB, grain number is denoted as LS, stalk weight is denoted as BZ, fruit weight is denoted as GZ, cluster weight is denoted as SZ, 1000-kernel weight is denoted as QLZ, longitudinal diameter is denoted as ZJ, transversal diameter is denoted as HJ, fruit shape index is denoted as GXZS, and fruit weight per fruit is denoted as GXZS. Fruit shape index is denoted as GXZS, single-fruit weight is denoted as DGZ, peroxidase is denoted as POD, superoxide dismutase is denoted as SOD, soluble protein is denoted as SP, and soluble sugar is denoted as SS (detailed data values are provided in Appendix A).

*2.6. Principal Component Analysis of I. polycarpa Fruit*

A principal component analysis of the 12 indicators from five *I. polycarpa* sources was followed by a dimensionality reduction analysis. Prior to the PCA, the 12 physiological parameters (e.g., soluble sugar, SOD activity) were standardized by autoscaling to account for differences in units and magnitude. The correlation matrix was used as an input for PCA to equalize the variable contributions. The results show that four principal components were extracted, with variance contribution rates of 52.1%, 24.3%, 8.2%, and 5.0%, respectively, and the cumulative contribution rate of the four principal components was 89.72%, transforming the original 12 indicators into four composite indicators (Table 4). As can be seen from Table 5, among the four main components, there are eight indicators with large loading values (>0.8): oil content, cluster length, number of grains, fruit weight, cluster weight, shank weight, fruit shape index, soluble protein, and POD. These can be used as key indicators to evaluate the quality of *I. polycarpa* fruit. Based on the data results in Table 5, quoting the coefficients in the table, four scores can be obtained.

$$F_k = \sum_{i=1}^{12} \omega_{ki} S_i, \quad (1)$$

$F_k$: kth principal component. $k = 1, 2, 3, 4$.

$\omega_{ki}$: the weight coefficient of variable $S_i$ in the kth principal component (i.e., the coefficient value in the original equation).

$S_i$: original variables (S1–S12).

**Table 4.** Eigenvalues and contribution values of the principal component analysis of fruit quality indices of different *I. polycarpa* ecotypes.

| Principal Components | Eigenvalues | Contribution Rate (%) | Cumulative Contribution Rate (%) |
|---|---|---|---|
| F1 | 6.258 | 52.1 | 52.15 |
| F2 | 2.921 | 24.3 | 76.49 |
| F3 | 0.987 | 8.2 | 84.71 |
| F4 | 0.601 | 5.0 | 89.72 |
| F5 | 0.455 | 3.8 | 93.51 |
| F6 | 0.303 | 2.5 | 96.03 |
| F7 | 0.191 | 1.6 | 97.63 |
| F8 | 0.121 | 1.0 | 98.63 |
| F9 | 0.081 | 0.7 | 99.31 |
| F10 | 0.064 | 0.5 | 99.85 |
| F11 | 0.018 | 0.2 | 100 |
| F12 | $5.524 \times 10^{-5}$ | 0 | 100 |

**Table 5.** Principal component analysis eigenvectors of fruit quality indexes of different *I. polycarpa* ecotypes.

| Indicator | F1 | F2 | F3 | F4 |
|---|---|---|---|---|
| Oil content | 0.47 | 0.813 | 0.04 | 0.098 |
| Cluster length | 0.924 | 0.048 | −0.163 | −0.112 |
| Cluster width | 0.787 | 0.317 | 0.065 | 0.311 |
| Number of fruits per cluster | 0.975 | −0.005 | −0.066 | 0.002 |
| Fresh weight of fruit stalks per cluster | 0.878 | 0.239 | −0.284 | −0.038 |
| Fruit weight per cluster | 0.931 | 0.172 | −0.106 | −0.032 |
| Cluster weight | 0.941 | 0.181 | −0.124 | −0.036 |
| Fruit shape index | −0.361 | −0.06 | 0.904 | −0.116 |
| POD | −0.161 | −0.909 | 0.036 | 0.147 |
| SOD | 0.141 | 0.697 | 0.366 | −0.452 |
| Soluble protein | 0.004 | 0.869 | −0.267 | −0.268 |
| Soluble sugar | −0.007 | −0.495 | −0.122 | 0.779 |

F1–F4 represent principal component 1, principal component 2, principal component 3, and principal component 4; S1–S12 represent the standardized values of the average of the 12 fruit indicators (oil content, cluster length, cluster width, number of grains, shank weight, etc., respectively); the four principal components and the corresponding variance contribution rates of the principal components are taken as the weights, to obtain the following comprehensive evaluation function:

$$F = 0.52148F1 + 0.24339F2 + 0.08221F3 + 0.05011F4, \quad (2)$$

According to the comprehensive evaluation function, the comprehensive scores of the evaluation of five ecotypes of *I. polycarpa* were calculated, respectively, and the higher scores indicated that the quality of the samples was better. Based on the results of the principal component analysis, the qualities of the five *I. polycarpa* ecotypes were found to be GG, SG2, HX, SH, and SG1, in descending quality order (Table 6).

**Table 6.** Comprehensive evaluation of the fruits of different *I. polycarpa* ecotypes.

| Seed Source | F1 | F2 | F3 | F4 | F (Full) | F_abridged | Rank (Full) |
|---|---|---|---|---|---|---|---|
| SH | 1.006 | 0.857 | 0.842 | −0.009 | 0.801 | 0.732 | 4 |
| SG1 | 0.701 | 0.419 | 0.335 | 0.223 | 0.496 | 0.467 | 5 |
| GG | 3.827 | 1.110 | −0.224 | 0.354 | 2.247 | 2.264 | 1 |
| HX | 2.090 | 1.005 | 0.113 | 0.088 | 1.343 | 1.333 | 3 |
| SG2 | 2.115 | 1.124 | −0.218 | 0.246 | 1.358 | 1.375 | 2 |

## 3. Discussion

Significant differences were observed among different ecotypes, with GG performing optimally in terms of yield and oil content, which may have a strong photosynthetic capacity and promote the accumulation of photosynthetic products. In contrast, the SOD activity and soluble protein content of HX were higher, suggesting that HX may have strong resistance to adversity and is suitable for promoting cultivation under adverse conditions. In terms of developmental stages, the fresh weight of single cluster and the oil content of fruits at stage M4 generally peaked, which was consistent with the pattern of dry matter and oil accumulation in the late stage of plant reproductive growth [29]. However, the sources from SG2 showed a high fruit shape index and fruit grain number at stage M1, which may be related to early maturity characteristics and potentially valuable for shortening the breeding time.

According to the results of the correlation analyses, oil content was significantly negatively correlated with water content, which was consistent with the competitive relationship of water metabolism during plant oil synthesis [30]. Meanwhile, oil content was positively correlated with yield traits such as cluster length and cluster width, suggesting that the better cultivars may have both high yield and high oil characteristics, which provides important guidance for the breeding of *I. polycarpa* as an oilseed crop. The significant negative correlations between POD activity, oil content, and SOD activity indicated that the antioxidant enzyme system may affect the stability of oils and fats. It has been shown that high POD activity may lead to lipid oxidation [31,32], which was confirmed by the higher oil content and lower POD activity of the seeds from GG in this study. In addition, SOD activity and soluble protein showed a significant positive correlation, indicating a possible synergistic regulation between the antioxidant system and protein synthesis, which was more evident in the HX seed source.

From the results of the principal component analysis, it can be seen that the cumulative contribution of the first four principal components reached 89.72%, which can reflect the main trait variations of *I. polycarpa*. Among them, principal component 1 mainly represented yield-related traits, which were consistent with the study of economic traits of *I. polycarpa* by Zhang Jian and Liu Ying [33,34]. Principal component 1 indicated that the yield traits dominant in the phenotypic variation of *I. polycarpa* and could be used as the core index for cultivar selection. Principal component 2 mainly represented the lipid and protein indices, which were positively correlated with oil content and soluble protein and negatively correlated with POD activity, indicating that the lipid accumulation of *I. polycarpa* might have an antagonistic relationship with antioxidant enzyme activity. This finding is consistent with those presented in the study of Jiang [31] in oilseed crops: high-oil-content cultivars may be accompanied by lower antioxidant enzyme activities to reduce the oxidative loss of oil. In contrast, the high negative correlation of POD alongside the positive correlation of soluble protein and the negative correlation of sugar content suggests a trade-off between oxidative defenses and carbon allocation. Principal component 3 mainly represents the fruit morphological characteristics. Principal component 4 showed close relations to soluble sugars, suggesting that sugar accumulation may be independent of oil accumulation and yield.

## 4. Materials and Methods

### 4.1. Overview of the Test Site

The experimental site was the Forestry Experimental Site, Science and Education Park, Henan Agricultural University, Zhengzhou City, Henan Province (113.6° E, 34.87° N; see the location map in Appendix A). The area has a typical semi-arid and semi-humid continental monsoon climate, with an average annual temperature of 14.2 °C, an average multi-year precipitation of 649.9 mm, and annual sunshine of about 2400 h. The soil is a sandy loam, with a pH value of 7.85, and a soil organic matter content of about 15 g/kg.

### 4.2. Test Material

Five ecotypes of seeds named after their origin were selected: Shanxi Hanzhong (SH), Sichuan Guangyuan1 (SG1), Guangdong Guangzhou (GG), Henan Xinyang (HX), and Sichuan Guangyuan2 (SG2), were sown at 2 m × 2 m spacing. The experimental material consisted of 5-year-old seedlings (3 for each ecotype), i.e., a total of 15 plants. On 25 June (M1), 25 July (M2), 25 August (M3), and 25 September (M4) 2024, three fruit clusters were randomly selected from the periphery of the crown of the target plants at the test site, and phenotypic and physiological indices of the fruits from each of the ecotypes in each month were determined. Three technical replications were performed for each indicator.

### 4.3. Test Methods

#### 4.3.1. Determination of Fruit Phenotypic Traits

Using a straight edge to determine the length and width of the clusters, we calculated the aspect ratio and determined the number of fruits per cluster. Then, we used an electronic balance to determine the seed weight per thousand grains, the fresh weight of the fruit stalk per cluster, the fruit weight per cluster, the single fruit fresh mass, and the fresh weight per cluster. Next, we used vernier calipers to determine the fruit width and fruit length and calculate the width–length ratios. Then, the thousand-kernel weight was obtained by randomly removing pure fruit kernels from the sample, counting 100 seeds with tweezers, weighing them and, multiplying this by 10 to obtain the thousand-kernel weight.

#### 4.3.2. Determination of Water and Oil Content of Fruit

The fruit water content ($\omega_w$) was determined as follows: 30 g of fresh fruits were weighed using a balance ($m_f$); they were put into the oven at 120 °C to bake for 0.5 h; then, the oven temperature was adjusted to 80 °C to dry them to constant weight ($m_d$); next, the water content was calculated using the following formula:

$$\omega_w = \frac{m_f - m_d}{m_f} \times 100\% \quad (3)$$

The fruit oil content was determined as follows ($\omega_o$): The Soxhlet extraction method was used to determine the oil content of the dried fruits. The dried fruit sample was put into the grinder. Then, 3 g was wrapped in filter paper and loaded into the Soxhlet extraction column. The appropriate amount of petroleum ether was put into the extraction bottle and extracted in a water bath for 10 h. Next, the filter paper was taken out and placed in the oven for drying to obtain the dry mass of the sample after extraction ($m_d$). The oil content was calculated as follows:

$$\omega_o = \frac{3 - m_d}{3} \times 100\% \quad (4)$$

#### 4.3.3. Determination of Fruit Contents and Antioxidant Enzymes

Soluble protein (SP) content was determined using the Coomassie Brilliant Blue G-250 staining method; soluble sugar (SS) content was determined using the anthrone method; peroxidase (POD) content was determined using the guaiacol method; superoxide dismutase (SOD) content was determined using the nitrogen blue tetrazolium photochemical reduction method [35]. SOD and POD indicate the amount of substrate that can be converted per minute, per gram of dry fruit weight.

SP was determined as follows: A measure of 0.5 g of the fresh sample was weighed; then, 5 mL of phosphate buffer was added. Next, following ice bath grinding, centrifugation was carried out at 4 °C 12,000× g for 15 min. Then, 100 µL of supernatant and 5 mL of Kaomas Brilliant Blue G-250 reagent were added and mixed in; this was left to stand for 10 min. The absorbance was measured at 595 nm with a spectrophotometer.

SS was determined as follows: A measure of 0.5 g of the fresh sample was added to 5 mL of distilled water; this was extracted in boiling water bath for 30 min, cooled, and then centrifuged at 8000× g at 4 °C for 10 min. Next, 0.5 mL of the supernatant was taken; then, 5 mL of anthrone reagent (0.2% anthrone in concentrated sulfuric acid solution) was added, and the sample was cooled in an ice bath after being in a boiling water bath for 10 min. The absorbance value was measured at 620 nm with a spectrophotometer.

SOD was determined by adding 1.5 mL of phosphate buffer (pH 7.8), 0.1 mL of Met solution, 0.1 mL of NBT solution, 0.1 mL of EDTA-$Na_2$ solution, 0.1 mL of riboflavin solution, and 0.1 mL of enzyme solution (the control was replaced by buffer) to a test tube. The reaction was placed under 4000-lux light for 20 min (25 °C), and the reaction was terminated by removing the light. Measure the absorbance value at 560 nm with a spectrophotometer.

POD was determined by taking 50 mL of phosphate buffer and adding 28 μL of guaiacol; this was heated and stirred before cooling, adding 19 μL of 30% $H_2O_2$ to form a reaction solution. Then, 3 mL of reaction solution and 1 mL of enzyme solution were added to a cuvette. The absorbance change was immediately recorded at 470 nm on a spectrophotometer reading every 30 s for 1 min.

### 4.4. Data Processing

The ANOVA and principal component analysis were performed using IBM SPSS Statistics 22. The correlation analysis and graphing were performed using Origin 2022. The location annotation was performed using ArcGIS 10.8.

## 5. Conclusions

In summary, the GG ecotype was found to be the preferred material for high-yield cultivation, especially at the M4 stage, when harvesting is economically optimal. The HX and SG1 ecotypes have the potential to be applied in quality improvement and specialty cultivation. Depending on the developmental characteristics of the different seed ecotypes, differentiated cultivation strategies can be developed. This study clarifies the key factors influencing the yield, quality, and stress-tolerance traits of *I. polycarpa* by analyzing *I. polycarpa* from multiple perspectives; thus, the results provide a theoretical basis for cultivar selection and breeding. In the future, we can combine transcriptomics or metabolomics to further resolve the molecular regulatory mechanisms of oil content and antioxidant enzyme activities, and we can carry out cultivation experiments in different ecological zones to optimize the cultivation strategy of *I. polycarpa*.

**Author Contributions:** Conceptualization, Z.L. (Zhi Li) and T.Z.; methodology, Y.Y., Q.Y., C.M., W.Z., C.C. and Z.H.; software, Y.Y., Q.Y., C.M., W.Z., C.C. and Z.H.; validation, Y.Y., Q.Y., C.M., W.Z., C.C. and Z.H.; formal analysis, Y.Y., Q.Y., C.M., W.Z., C.C. and Z.H.; investigation, Y.Y., Q.Y., C.M., W.Z., C.C. and Z.H.; resources, Y.Y., Q.Y., C.M., W.Z., C.C. and Z.H.; data curation, Y.Y., Q.Y., C.M., W.Z., C.C. and Z.H.; writing—original draft preparation, Y.Y.; writing—review and editing, Z.L. (Zhen Liu), Y.W., X.G., Q.C., H.L., L.D., J.W., Y.R. and F.L. visualization, Y.Y.; supervision, Z.L. (Zhi Li) and T.Z.; project administration, Z.L. (Zhi Li) and T.Z.; funding acquisition, Z.L. (Zhi Li) and T.Z. All authors have read and agreed to the published version of the manuscript.

**Funding:** This research was funded by the Postgraduate Education Reform and Quality Improvement Project of Henan Province (YJS2024AL062/YJS2025AL63), the Henan Provincial Social Sciences Federation Research Project in China (SKL-2024-2179), the Henan Agricultural University's "Millions of Science and Education Service Action" Fund Project (2024SFBQW30), and the Henan Province Postdoctoral Research Project of China (202002053). Zhejiang Shuren University Basic Scientific Research Special Funds (2025XZ005). Key Specialty Project of Ordinary Colleges and Universities in Shaoxing City, Zhejiang Province, "Landscape Architecture" (SXSZY202412).

**Data Availability Statement:** The datasets generated and/or analyzed during the current study are available from the corresponding author upon reasonable request.

**Acknowledgments:** We are grateful to the reviewers for their insightful comments, which contributed to the quality of the article. We would also like to thank our laboratory team, especially Sasha Wang, who assisted with the research.

**Conflicts of Interest:** The authors declare no conflicts of interest.

## Appendix A

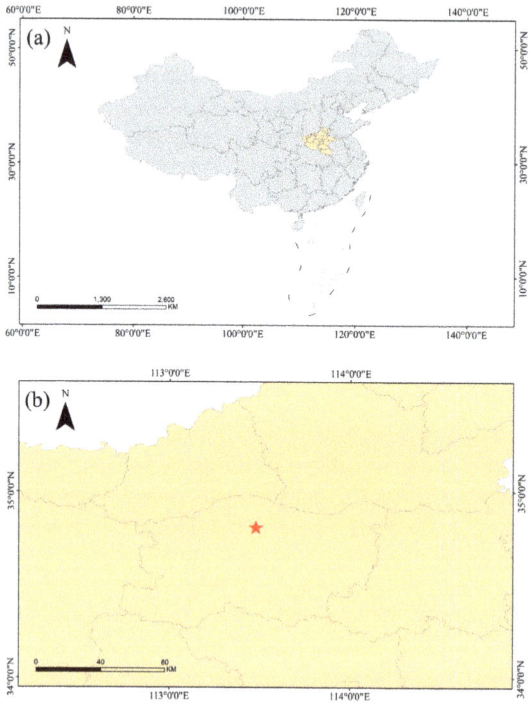


**Figure A1.** Geographical location of the test site. (**a**) The map of China and the Henan province (yellow areas) where the study was conducted. (**b**) The specific location of the test site (red star).

| | HSL | HYL | GSC | GSK | GSCKB | LS | BZ | GZ | SZ | QLZ | ZJ | HJ | GXZS | DGZ | POD | SOD | SP | SS |
|---|---|---|---|---|---|---|---|---|---|---|---|---|---|---|---|---|---|---|
| HSL | 1 | | | | | | | | | | | | | | | | | |
| HYL | −0.630** | 1 | | | | | | | | | | | | | | | | |
| GSC | −0.137 | 0.404** | 1 | | | | | | | | | | | | | | | |
| GSK | −0.222 | 0.481** | 0.634** | 1 | | | | | | | | | | | | | | |
| GSCKB | 0.027 | 0.126 | 0.738** | −0.043 | 1 | | | | | | | | | | | | | |
| LS | −0.106 | 0.401** | 0.848** | 0.678** | 0.488** | 1 | | | | | | | | | | | | |
| BZ | −0.215 | 0.485** | 0.802** | 0.618** | 0.501** | 0.882** | 1 | | | | | | | | | | | |
| GZ | −0.163 | 0.522** | 0.764** | 0.664** | 0.404** | 0.888** | 0.848** | 1 | | | | | | | | | | |
| SZ | −0.17 | 0.525** | 0.778** | 0.668** | 0.419** | 0.900** | 0.875** | 0.999** | 1 | | | | | | | | | |
| QLZ | 0.074 | −0.038 | 0.071 | 0.127 | −0.036 | −0.016 | −0.057 | 0.188 | 0.167 | 1 | | | | | | | | |
| ZJ | −0.074 | 0.09 | −0.279* | −0.163 | −0.214 | −0.399** | −0.274* | −0.015 | −0.041 | 0.478** | 1 | | | | | | | |
| HJ | −0.038 | 0.204 | −0.018 | −0.002 | −0.003 | −0.15 | 0.043 | 0.244 | 0.227 | 0.490** | 0.823** | 1 | | | | | | |
| GXZS | −0.063 | −0.195 | −0.404** | −0.254 | −0.321* | −0.368** | −0.494** | −0.428** | −0.441** | −0.109 | 0.175 | −0.413** | 1 | | | | | |
| DGZ | −0.087 | 0.16 | −0.128 | −0.075 | −0.086 | −0.286* | −0.099 | 0.118 | 0.098 | 0.512** | 0.911** | 0.965** | −0.212 | 1 | | | | |
| POD | 0.654** | −0.633** | −0.166 | −0.256* | 0.014 | −0.155 | −0.335** | −0.315* | −0.321* | −0.076 | −0.312* | −0.357** | 0.112 | −0.371** | 1 | | | |
| SOD | −0.566** | 0.479** | 0.064 | 0.062 | 0.022 | 0.071 | 0.153 | 0.163 | 0.164 | −0.051 | 0.285* | 0.138 | 0.228 | 0.199 | −0.617** | 1 | | |
| SP | −0.383** | 0.598** | 0.16 | 0.153 | 0.081 | 0.039 | 0.275* | 0.113 | 0.131 | −0.078 | 0.176 | 0.283* | −0.201 | 0.279* | −0.725** | 0.515** | 1 | |
| SS | 0.527** | −.360** | −0.096 | −0.035 | −0.101 | 0.015 | −0.059 | −0.091 | −0.099 | −0.178 | −0.359** | −0.25 | −0.157 | −0.317* | 0.490** | −0.563** | −0.521** | 1 |

* P<0.05 ** P<0.01

**Figure A2.** Correlation analysis of fruit quality indexes of the different ecotypes of *I. polycarpa*. Water content is denoted as HSL, oil content is denoted as HYL, cluster length is denoted as GSC, cluster width is denoted as GSK, cluster aspect ratio is denoted as GSCKB, grain number is denoted as LS, stalk weight is denoted as BZ, fruit weight is denoted as GZ, cluster weight is denoted as SZ, 1000-kernel weight is denoted as QLZ, longitudinal diameter is denoted as ZJ, transversal diameter is denoted as HJ, fruit shape index is denoted as GXZS, and fruit weight per fruit is denoted as GXZS. Fruit shape index is denoted as GXZS, single-fruit weight is denoted as DGZ, peroxidase is denoted as POD, superoxide dismutase is denoted as SOD, soluble protein is denoted as SP, soluble sugar is denoted as SS.

## References


1. Zhu, L.; Guo, X.H. Cultivation technology of *Idesia polycarpa*, an excellent woody oilseed tree species. *Shaanxi For.* **2008**, *5*, 34.
2. Li, X. Current situation and prospect of oil grape (*Idesia polycarpa*) industry in China. *China For. Ind.* **2019**, *2*, 68–73.
3. Cai, Q.F.; Wang, S.; Xu, Z.H.; Song, Y.; Liu, Z. Analysis of the differences in chlorophyll fluorescence characteristics between different seed sources of *Idesia polycarpa*. *Henan Sci.* **2021**, *39*, 715–719.
4. Tian, X.X.; Fang, X.Z.; Du, M.H. Nutritional quality and antioxidant capacity of oil from different parts of *Idesia polycarpa* fruit. *Chin. J. Cereals Oils* **2020**, *35*, 91–95.
5. Jiang, H.Y. Preparation of Conjugated Linoleic Acid and Edible Oil Processes from *Idesia polycarpa* Oil. Master's Thesis, Chengdu University, Chengdu, China, 2023.
6. Jiang, H.Y.; Pang, G.Y.; Guo, X.Q.; Chen, L.M.; Kang, T.R. Response surface optimisation for the preparation of conjugated linoleic acid from *Idesia polycarpa* oil. *China Flavour.* **2022**, *47*, 92–98.
7. Wang, Q.Y.; Du, K.F.; Jia, C.M.; Li, X.; Yao, S.; Song, H. Optimization of Linoleic Acid from *Idesia polycarpa* Maxim. var. vestita Diels by Response Surface Methodology. *Adv. Mater. Res.* **2011**, *396–398*, 1222–1227. [CrossRef]
8. Liu, D.J.; Yan, P. Analysis of the industrialisation system and management strategy of China's bio-industrial tree species *Idesia polycarpa*. *For. Resour. Manag.* **2015**, *2*, 9–12.
9. Yang, F.X.; Su, Y.Q.; Li, X.H.; Zhang, Q.; Sun, R.C. Preparation of biodiesel from *Idesia polycarpa* var. vestita fruit oil. *Int. News Fats Oils Relat. Mater. INFORM* **2009**, *20*, 320.
10. Zhu, Z.Y.; Wang, Q.; Ruan, X.; Li, Z.H.; Xue, J.; Jiang, H.; Lu, X. Oil content and fatty acid content of fruits from different geographic populations of *Idesia polycarpa*. *For. Sci.* **2010**, *46*, 176–180.
11. Wu, L.L.; Deng, W.F.; Lu, X.Y.; Niu, C.T.; Tian, H.; Li, Z. Progress of research on the development and utilisation of *Idesia polycarpa*. *Res. Econ. For.* **2023**, *41*, 242–252.
12. Wang, Y.M.; Wang, D.H.; Liu, Z.; Wang, L.; Jing, Z.J.; Dai, L. GC/MS analyses of oil content and fatty acids in seeds of *Idesia polycarpa* from different seed sources. *Econ. For. Res.* **2011**, *29*, 14–21.
13. Jiang, X.B.; Li, D.W.; Gong, B.C.; Wu, K.Y. A preliminary study on seedling growth pattern of *Idesia polycarpa* from different origins. *Henan Agric. Sci.* **2012**, *41*, 133–136.
14. Gong, B.C.; Li, D.W.; Jiang, X.B.; Wu, K.Y.; Peng, J.L.; Bai, J.J. Analysis of variations in fatty acid composition of fruits from different seed sources of *Idesia polycarpa*. *J. Plant Physiol.* **2012**, *48*, 505–510.
15. Hernández, M.L.; Sicardo, M.D.; Belaj, A.; Martínez-Rivas, J.M. The Oleic/Linoleic Acid Ratio in Olive (*Olea europaea* L.) Fruit Mesocarp Is Mainly Controlled by OeFAD2-2 and OeFAD2-5 Genes Together With the Different Specificity of Extraplastidial Acyltransferase Enzymes. *Front. Plant Sci.* **2021**, *12*, 653997. [CrossRef]
16. Zhang, W.L.; Zhao, C.W.; Karrar, E.; Du, M.; Jin, Q.Z.; Wang, X.G. Analysis of Chemical Composition and Antioxidant Activity of *Idesia polycarpa* Pulp Oil from Five Regions in China. *Foods* **2023**, *12*, 1251. [CrossRef]
17. Ye, Y.; Tang, X.S.; Chen, F.; Tang, L. Optimization of Phenolics Extracted from *Idesia polycarpa* Defatted Fruit Residue and Its Antioxidant and Depigmenting Activity In Vitro and In Vivo. *Evid.-Based Complement. Altern. Med. eCAM* **2014**, *2014*, 931269. [CrossRef]
18. Xu, Z.H.; Fang, L.S.; Liu, Z.; Li, Z.; Cai, Q.F.; Geng, X.D.; Zhai, W.J.; Wang, B.Y.; Zhan, Z.W.; Wang, Y.M. Analysis of photosynthetic characteristics of different seed sources of *Idesia polycarpa*. *J. Henan Agric. Univ.* **2021**, *55*, 44–51.
19. Zhou, K.; Chang, Y.H.; Song, M.F.; Liu, X.Y.; He, J.; Ma, L.Z. Quality analysis and comprehensive evaluation of *Idesia polycarpa* from different origins. *China Oil Fats* **2025**, *50*, 129–134+152.
20. Xu, S. Growth performance and stress tolerance of Sanguisorba officinalis under multi-environmental conditions. *For. Sci. Technol. Intell.* **2025**, *57*, 53–55.
21. Geng, X.D.; Yue, J.F.; Liu, Y.; Tao, Y.H.; Zhou, Y.F.; Zhang, Z.Z.; Xie, H.D.; Cai, Q.F.; Liu, Z. Response of "Yuji" shantung seedling growth to drought stress. *Econ. For. Res.* **2025**, *43*, 29–39.
22. Geng, X.D.; Zhou, Y.F.; Liu, Y.; Tao, Y.H.; Xie, H.D.; Yue, J.F.; Zhang, Z.Z.; Cai, Q.F.; Dai, L.; Liu, Z. Effects of flooding stress on growth and physiology of "Yuji" shantung seed. *J. Henan Agric. Univ.* **2025**, *59*, 389–399.
23. Tian, K.X.; Xu, J.Y.; Dai, L.; Li, Z.; Geng, X.D.; Liu, Z.; Wang, Y.M. Physiological response of *Idesia polycarpa* reedlings to extreme high temperature and high temperature plus drought stress. *For. Sci.* **2024**, *60*, 84–92.
24. Feng, J.; Yuan, Q.P.; Chen, X.Z.; Fang, L.S.; Zhang, T.; Liu, Z.; Wang, Y.M.; Geng, X.D.; Cai, Q.F.; Li, Z. Identification and Chemical Control of Stem Canker Pathogen of *Idesia polycarpa*. *Plants* **2025**, *14*, 1393. [CrossRef]
25. Li, Z.; Yang, Y.; Feng, J.; Rana, S.; Wang, S.S.; Wang, H.M.; Zhang, T.; Wang, Y.M.; Guo, G.P.; Cai, Q.F.; et al. Evaluation of Fine Root Morphology and Rhizosphere Environmental Characteristics of the Dioecious *Idesia polycarpa* Maxim. *Forests* **2024**, *15*, 234. [CrossRef]
26. Zhang, M.X.; Rana, S.; Li, C.Z.; Zhang, X.X.; Tian, K.X.; Liu, Z.; Li, Z.; Dai, L.; Geng, X.D.; Wang, Y.M. Transcriptome analysis reveals the role of temperature in seed germination of *Idesia polycarpa* Maxim through the integration of phytohormones and sugar metabolism. *Braz. J. Bot.* **2024**, *47*, 963–979. [CrossRef]

27. Wang, S.S.; Rana, S.; Zhang, T.; Wang, Y.M.; Liu, Z.; Cai, Q.F.; Geng, X.D.; Yuan, Q.P.; Yang, Y.; Miao, C.; et al. Influence of Varied Phosphorus Fertilizer Ratios on the Rhizosphere Soil Microbial Community in *Idesia polycarpa* Seedlings. *Forests* **2024**, *15*, 1686. [CrossRef]
28. Wang, H.M.; Li, Z.; Cai, Q.F.; Wang, Y.M.; Geng, X.D.; Li, S.F.; Fang, L.S.; Yao, S.Y.; Li, H.Y.; Liu, Z. Mutant Bisexual and Wild Male Flowers Were Compared by Integrated Proteome and Transcriptome Analyses to Provide Insight into the Sex Conversion of *Idesia polycarpa* Maxim. *Forests* **2023**, *14*, 1737. [CrossRef]
29. Xie, H.D.; Xu, Z.H.; Tao, Y.H.; Zhou, Y.F.; Yue, J.F.; Zhang, Z.Z.; Geng, X.D.; Cai, Q.F.; Liu, Z. Changes in oil accumulation and nutrient content during fruit development of *Idesia polycarpa*. *Econ. For. Res.* **2025**, *43*, 94–103.
30. Cinosi, N.; Moriconi, F.; Farinelli, D.; Marchionni, D.; Lodolini, E.M.; Rosati, A.; Famiani, F. Effects of summer pruning on the water status and physiology of olive trees and on fruit characteristics and oil quality. *Sci. Hortic.* **2024**, *324*, 112612. [CrossRef]
31. Jiang, S.Q. Effects of Soybean Isoflavones on Production Performance, Meat Quality and Antioxidant Mechanism of Lingnan Yellow-Feathered Broilers. Doctoral Thesis, Zhejiang University, Hangzhou, China, 2007.
32. Xiang, X.W.; Wen, L.Y.; Wang, Z.R.; Yang, G.; Mao, J.M.; An, X.F.; Kan, J.Q. A comprehensive study on physicochemical properties, bioactive compounds, and emulsified lipid digestion characteristics of *Idesia polycarpa* var. Vestita Diels fruits oil. *Food Chem.* **2023**, *404*, 134634. [CrossRef]
33. Zhang, J.; Wu, R.R.; Guo, G.Z.; Deng, T.; Liang, Y.S. Evaluating Phenotypic Traits of Fruits on Single Tree of *Idesia polycarpa* in Southwestern Guizhou. *Guizhou For. Sci. Technol.* **2025**, *53*, 65–72.
34. Liu, Y.; Wang, S.; Wang, H.M.; Liu, Z. Variation patterns of fruit traits and selection of good single plants in *Idesia polycarpa*. *J. For. Environ.* **2022**, *42*, 536–543.
35. Cai, Q.S. *Plant Physiology Experiment*; China Agricultural University Press: Beijing, China, 2013; pp. 74–78.

MDPI AG
Grosspeteranlage 5
4052 Basel
Switzerland
Tel.: +41 61 683 77 34

*Plants* Editorial Office
E-mail: plants@mdpi.com
www.mdpi.com/journal/plants

www.ingramcontent.com/pod-product-compliance
Lightning Source LLC
Chambersburg PA
CBHW062053220926
51850CB00010B/2757
* 9 7 8 3 7 2 5 8 7 5 4 4 3 *